The Asian Hornet Handbook

First published
2019
Psocid Press

ISBN-13
978-1-9160871-0-1

 Sarah Bunker CC-BY-SA 4.0

This work is licensed under the Creative Commons Attribution-ShareAlike 4.0 International License. To view a copy of this license, visit http://creativecommons.org/licenses/by-sa/4.0/ or send a letter to Creative Commons, PO Box 1866, Mountain View, CA 94042, USA.

Thank you to those whose CC images I have used from Wikipedia: you are properly credited at the photos, and your licenses can be found at the Creative Commons website above.

Psocid (pronounced 'so-sid') Press runs on tea and jaffa cakes.

For Paul

Contents

Introduction. 1

Biology

Asian Hornet Identification. 3
Asian hornet biology. 11
  Life history 11
    Emergence from hibernation. 11
    Founding — the queen builds a primary nest 12
    Producing the first workers 14
    Early dangers 16
    The primary nest grows 17
    Relocation 19
    The secondary nest 19
    Foraging 25
    Food for adults 27
    Food for larvae 27
    Honey bee predation. 30
    Honey bee defence. 33
    Production of the sexual stages 36
    Mating 38
    Nest decline 39
    Hibernation of the queens 39
  Ecology 39
    Problems with invasive alien species 39
    Effect on honey bees. 40
    Effect on European hornets 41
    Competition among Asian hornets. 42
    Effects on humans 42
  Asian Hornet Stings 43

Context

The spread of the Asian hornet. 47
  The Asian hornet arrives in Europe 47
  Asian hornets reach the Channel Islands 50
  Asian hornets in the UK. 54
A National Contingency Plan 59
  The main players 59
  Waiting...and spotting. 61
  Dealing with an isolated incident 63
  Losing control. 64
  The 'new normal' 64
The birth of Asian hornet action teams. 65

Control

Finding Asian hornet nests 69
  Introduction. 69
  The Jersey method 70

    Radio telemetry 90
    Other methods of finding nests 93

Asian hornet colony destruction. 95

Trapping. 101
    Context 101
    Monitoring traps. 101
    When does a monitoring trap become a
    killing trap? 102
    Monitoring without trapping. 102
    Trapping is contentious 104
    The effect of trapping on non-target insects:
    the data from France 104
    Does spring trapping reduce the number of Asian
    hornet colonies?. 105
    Usurpation — natural population control? 107
    UNAF on trapping 108
    Where others stand in France 108
    Conclusions from French spring trapping data. 109
    Trap practicalities 110
    Diversion traps 110
    Trap design 110
    Raoul's cage 115
    The JABEPRODE 115
    The ApiShield 116
    Type of attractant 118
    Trapping at different times has different results. 119
    Trapping conclusions 122

Poisoning. 123

Biological control. 125
    Conops vesicularis 125
    Other insect parasites 126
    Nematodes. 126
    Entomopathogenic fungi 127
    Carnivorous plants 127
    Chickens. 128
    Deformed wing virus. 128
    Mites. 128

Genetics. 129
    Selection of honey bee strains 129
    DNA technology 129

Hive defence. 131
    Reduction of entrances. 131
    Disrupting hawking 131
    Thwack and zap 133
    Feeding. 134
    Glossary. 135
    Resources. 138
    AHAT kit list 139
    References 140
    Acknowledgements 149
    Index 150

Introduction

Around 2004, an accidental import, possibly a single mated female, introduced the Asian or yellow-legged hornet (*Vespa velutina*) into southern France from China, and it has spread through western European countries like wildfire. Since 2016 there have been incursions into the UK, and if all are not dealt with they could establish quickly. Asian hornets cause two important problems for humans: they love eating bees, especially honey bees; and if a nest is accidentally disturbed they can be very aggressive.

In the summer of 2018, a call went out from Jersey beekeepers who needed help in tracking Asian hornets to their nests so that these could be destroyed. I travelled to Jersey with some other beekeepers from Devon and Somerset to help. Working in small teams, we were able to find about a nest a day, using very simple equipment. When I got back to England, I reflected on what I had learned and was determined to get people back home fully informed and ready to track if the need arises: I wanted us to be able to hit the ground running, and that is why I have written this handbook.

To stop Asian hornets from establishing in the UK, the first action must be to raise awareness; to inform people and teach them to recognise this insect. To this end, Asian Hornet Action Teams (AHATs) are being formed across the land by beekeepers who are educating people and helping the National Bee Unit narrow its response to actual sightings, sifted from thousands of misidentified insects.

This book is divided into three parts: it starts with **Biology**, looking in detail at what we know so far of their life history and ecology. I give no apology for the depth of this section; only by understanding their natural history can we hope to control them without upsetting the ecological balance. The **Context** gives information on the story as it has unfolded in Europe (including the UK); and in **Control**, tracking, destruction of nests and other methods are discussed.

I have tried to keep the language easy to follow, but there are quite a few specific terms that are important, so I have provided a Glossary (with an illustration of Asian hornet body parts for those not at all familiar with insects). For those who want to go deeper, there is a full reference list (most articles can be found online), a page of resources, and an AHAT kit list. There's also an index.

Sarah Bunker, March 2019

Part I
Biology

Asian Hornet Identification

Asian hornets are straightforward to identify, especially with their completely dark thoraxes and yellow legs; however, despite this and all the available information online and in print, in France 30% of the public still misidentify them (Rome et al. 2011a). The rate is probably worse in the UK, since we are still unfamiliar with this hornet. This is a major headache for the National Bee Unit (NBU) and the Centre for Ecology and Hydrology (CEH) who have to triage thousands of sightings to find the few that are actual Asian hornets.

The scientific species name for the Asian or yellow-legged hornet is *Vespa velutina*. In fact the full name for the subspecies that has invaded Europe is *Vespa velutina nigrithorax*: the last name refers to the black thorax. In Asia, where this hornet comes from, there are 12 subspecies, all with slightly different colouring. *Vespa velutina nigrithorax* is from Central and Eastern China.

It ranges from around 25 mm to 30 mm in length (head to tail). At the beginning of the season the workers are smaller, and those that are produced later in the year are larger, due to being better fed as larvae. Based just on size, it is impossible to tell queens apart from males or the female workers (although apparently this is possible in other subspecies, see Wen et al. 2017). The Asian hornet is mainly a velvety black or very dark brown, with a thin yellow band on the abdomen close to the thorax and a wider orange-yellow band closer to its tail on the abdomen. It also has an orange face (the head is black from above) and dark smoky-brown wings (which, like other hornets and wasps, are folded longitudinally into narrow strips and form a 'V' shape when it is not flying: see illustration). The top halves of its legs are a dark brown-black and the bottom halves are yellow (hence 'yellow-legged hornet'). It has long brown/black antennae. From underneath (if you see one in a transparent container), they have a bit more yellow and are not as distinctive (see illustration).

They only fly in full daylight (unlike the European hornet).

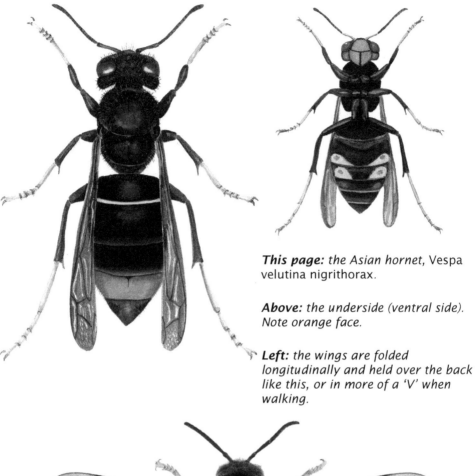

This page: *the Asian hornet,* Vespa velutina nigrithorax.

Above: *the underside (ventral side). Note orange face.*

Left: *the wings are folded longitudinally and held over the back like this, or in more of a 'V' when walking.*

Dried specimen. With its wings stretched out you can see that there are two pairs. The sting is also visible. The colours of the dried specimen are duller than in a live insect. Photo by Didier Descouens (CC BY-SA 4.0)

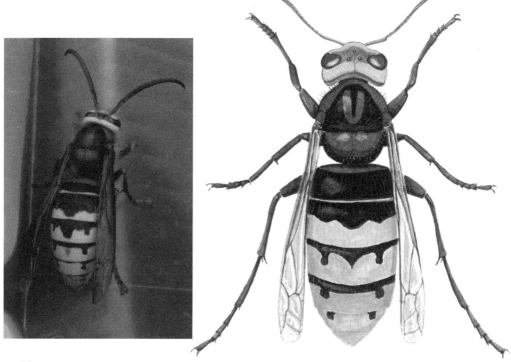

This page: our native European hornet, Vespa crabro.

What can Asian hornets be confused with?

European hornet (*Vespa crabro*)

Our native European hornet, *Vespa crabro*, is larger than *Vespa velutina* (it is 30-35 mm long) and looks like a very large wasp. It is the only other hornet seen in northern Europe. *Vespa crabro* is bulkier than the Asian hornet, with a large and sometimes drooping abdomen. The markings on its abdomen are a deep yellow, and black to chestnut brown; the pattern is very distinctive: the darker marks form shapes that look just like the lugs on jigsaw pieces. It also has chestnut (red-brown) legs and a black and brown thorax. It has a yellow face and brown to chestnut antennae. The native European hornet is often active at night, sometimes flying into rooms through open windows on warm evenings, attracted to the light.

Although our native hornet is big, it is no more likely to sting people than an ordinary wasp, but with its longer sting and larger size, it can deliver more venom and a painful sting.

Vespa crabro builds quite large paper nests, but they are usually under cover in a tree-hole, wall or loft space; occasionally underground. Although the colony (apart from the queen who flies off to hibernate) dies off in the autumn, the nest site may be used again the following year, unlike Asian hornets that apparently never re-use a nest. Native European hornet nests have a large entrance at the bottom, whereas only very small young nests of Asian hornets have an entrance at the bottom. In large (later season) Asian hornet nests, the entrance is made halfway up the side, and it is only around 2-3 cm across.

Horntail or Giant wood wasp (*Urocerus gigas*)

This spectacular insect is a type of sawfly. The females have what look like terrifyingly long stings, but these are in fact harmless ovipositors that they use to deposit their eggs in tree wood. They are quite a bit larger than Asian hornets — the females get up to 45 mm long. They have a distinctive yellow and black banded abdomen, yellow legs and long yellow antennae. They have yellow/golden patches behind their black eyes. They are found in woods and sometimes emerge from construction timber.

Urocerus gigas, *the horntail. Photo by Holger Gröschl (CC BY-SA 2.0)*

Hornet mimic hover-fly (*Volucella zonaria*)

Volucella zonaria, *hornet mimic hover-fly. Photo by Alvesgaspar (CC BY-SA 3.0)*

As the name suggests, this is a type of fly that has evolved to look dangerous like a hornet, but this is a ruse — it has no sting and is completely harmless. They are generally a bit smaller (20 mm or less) than an Asian hornet. The banding on the abdomen has much more yellow than the Asian hornet and its legs are dark. When it is walking, it puts its wings over its back, but they aren't folded up like a wasp or hornet. The most distinctive difference between *Volucella zonaria* and the Asian hornet is the antennae: the hover-fly has very small whisker-like antennae, whereas hornets have large distinctive antennae.

Wasps

Technically, hornets are wasps: to entomologists, they are part of a bigger 'wasp' family. What we generally refer to as wasps in the British Isles are the smaller black and yellow insects that can be annoying in pub gardens and orchards. They are significantly smaller than the Asian hornet (they are between 10 and 20 mm), and shinier. There are several species in this country; some build nests in protected, dark places, such as in the ground, in

Vespula vulgaris *(male). Photo by Magne Flåten (CC BY-SA 3.0)*

compost heaps and in building cavities, while others build exposed nests in bushes and small trees. The paper nest is roughly spherical and has a small entrance at the bottom. One thing that all these wasps have in common is some form of yellow marking on the thorax, whether spots or 'shoulder' stripes: the Asian hornet has a completely black thorax.

Dolichovespula media. *Photo by PJT56 / Wikimedia Commons / CC BY-SA 4.0*

Dolichovespula saxonica. *Photo by Sandy Rae (CC BY-SA 3.0)*

Vespula rufa. *Photo by Richard Bartz, Munich aka Makro Freak (CC BY-SA 2.5)*

A note about the Asian giant hornet

The Asian (or Japanese) **giant** hornet (*Vespa mandarinia*) has not been found in Europe, but that doesn't stop newspapers and online articles using pictures of it when they should be using pictures of *Vespa velutina*. This is the world's largest hornet (around 45 mm and chunky) and its size and completely yellow head make it more dramatic than *Vespa velutina*.

Vespa mandarinia japonica, *the Asian* **giant** *hornet (female). Photo by Yasunori Koide (CC BY-SA 4.0)*

Reporting a sighting of an Asian hornet

There is an app available for Apple and Android phones called 'Asian Hornet Watch' for identifying and reporting this hornet. Or you can send sightings with a photograph and location details to:

alertnonnative@ceh.ac.uk

The Non-Native Species Secretariat (NNSS) has produced some great identification posters. Keep one in your car or, if you are a beekeeper, pin one to the underside of one of your hive roofs for quick guidance.

Identifying different castes and sexes

Females (workers and queens) have 10 segments to their antennae, the males have 11 segments (the segments are difficult to count), and their antennae are usually more curved. Males also have blunter 'tails', and on their underside, close to the tip of the tail are two small yellow spots (see photo). Males don't have stings.

*Male Asian hornet showing two yellow spots (**arrow**) which distinguish it from a female. Photo by John de Carteret*

You are most likely to come across workers (from July onwards), because they are busy with colony tasks including foraging. They are present in the colony all through the season, once the queen has raised the first batch of them. Because they are female, they have stings.

Distinguishing queens from workers can be difficult. If you see an Asian hornet between the middle of February and 1st May (timings from France — it may be later if they become established in the UK), it is most likely to be a queen, because only the queen hibernates over winter (the other castes die off), and once she builds a nest it still takes another 50 days for the first workers to appear. Once the queen has spent a long time confined to the nest, she becomes shiny from constant attention by workers (Martin 2017). Queens are heavier than workers (Rome et al. 2015), but it is impossible to judge by eye because they don't appear any bigger; and to weigh them you need a seriously sophisticated balance. Probably the most accessible technique for distinguishing them is to measure the mesoscutum (the middle part of the thorax, see illustration). It has been established that workers have a mesoscutum width under 4.5 mm and queens have wider mesoscuta (Pérez-de-Heredia et al. 2017).

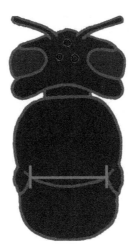

*The mesoscutum width (**pink line**).*

Asian Hornet Biology

Sarah Bunker and Lynne Ingram

Life history

Emergence from hibernation

The cycle starts when mated queens, also known as 'foundresses', come out of hibernation in the spring. They will have been hibernating in woodpiles, burrows, crevices, tree bark, soil or under stones (Monceau et al. 2014a, Marris et al. 2011) and may hibernate in clusters of two or three (Chauzat & Martin 2009), in a characteristic pose with the wings tucked under the abdomen (Martin 2017). In France, beekeepers expect foundresses to be on the wing once the temperature is consistently above 13 °C (AAAFAa 2015): certainly, none were trapped below 10 °C (Monceau et al. 2012). In the UK, this would translate to them appearing in March/April (but bear in mind climate change).

Once emerged, foundress queens rebuild their strength by feeding on nectar or tree sap. The most important food source for large-sized *Vespa* species in spring is the sap secreted from the trunks of oak in addition to those of beech, maple and willow (Matsuura & Yamane 1990). In Jersey and France, Asian hornet queens have also been noticed feeding on spring-flowering camellias (AAAFAb 2016).

In other species of *Vespa*, post-hibernation migration of queens en masse has been witnessed as swarms, or assumed from mark and recapture experiments when many marked queens have disappeared from an area (Matsuura & Yamane 1990). In *Vespa velutina*, it is assumed that some post-hibernation migration occurs, to account for the fast spread of this hornet through Europe (around 78 km per year, Robinet et al. 2016) but it is not known whether queens travel alone or in small groups. In unpublished data, Rome et al. (2015) found, in preliminary flight-mill experiments, that *Vespa velutina* queens are able to fly more than 40 km per day, although flight mills are a long way from natural flight (see 'Foraging' later in this section).

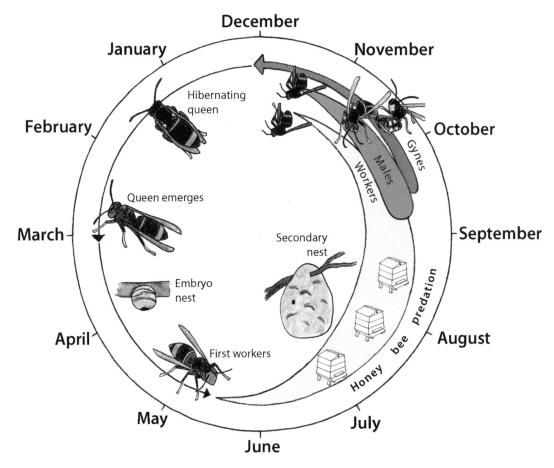

Life cycle of Vespa velutina. *The **yellow area** indicates the increase in numbers of workers. Towards the end of the season, the sexual stages are produced; first the males (**blue**) and then the females (**red**). The female sexual stages are called gynes and once fertilised will become next year's queens*

Founding — the queen builds a primary nest

During the feeding period, the queen's ovaries begin to develop and, as this happens, the nesting drive intensifies until the queen selects a nest site (Spradbery 1973).

Asian hornet nests are built, like those of other wasps and hornets, out of a kind of papier-mâché, made from harvesting plant material and pulping it with saliva and water. The papier mâché is then shaped into the structure of the nest, using the mouthparts. Because the plant material is from diverse sources, the colour of the material varies and results in coloured lines (browns, creams) that show how the material was laid down in thin strips, extending the edge of the paper. The overall effect is usually a pale brown or beige-looking nest. Because water is so important for nest building, and the fact that the nest is constantly being expanded, Asian hornets will often choose sites with good access to water in drier climates (Monceau et al. 2012); if there is water everywhere, this factor becomes less important (for example in Galicia, Rojas-Nossa et al. 2018).

In the spring the overwintered queen finds a location for her primary nest (in 2017, the first Jersey primary nest was reported on 18th April, and in 2018, it was on 23rd April, at which point the nests had 8-10 cells and the first eggs had been laid). This is built in a sheltered place as it is small, fragile and designed to just house her first batch of workers. The initial 'embryo' nest built solely by the foundress is only 4-5 cm across (Martin 2017); once the first workers emerge and start to enlarge it, it is called a primary nest. These nests are usually built at a height below 10 m, perhaps reducing the energy requirements of the queen (it takes a lot of effort to fly up to a high nest carrying pulp or prey, Rome et al. 2015). Humans build all sorts of structures which create ideal shelter for this first nest, including barns, sheds, garages and eaves. Foundresses also like enclosed spaces like holes in walls or trees; a bird box was even chosen in Jersey. In south-western France (Andernos-les-Bains), 30 of 39 primary nests (77%) were associated with such man-made structures (Franklin et al. 2017). These are the places to keep an eye on in the spring. In Jersey, primary nests were all found in or on man-made structures, including sheds, garages and porches, and all by members of the public.

The foundress starts the nest by making a stalk or petiole that extends down from a convenient beam or ceiling. The cells are built at the bottom of this stalk and a thin, spherical envelope, one or two layers thick, is built around the whole lot to protect the eggs and larvae and keep them warm. The resulting nest is like a ball, with the entrance at the bottom. Unlike honey bees, wasps and hornets build paper comb (the cells in which they raise their brood), which is horizontal rather than vertical, with the openings of the hexagonal cells pointing down.

The foundress is curled around the petiole, providing warmth to the eggs.
Photo by John de Carteret

Left: View looking up into embryo nest. The queen is again curled around the petiole. **Right**: Embryo nest attached to shed roof.
Both photos by John de Carteret

The queen is alone and vulnerable until the first workers emerge, and during this 'queen colony phase', she hunts for food for herself and to feed the larvae, lays eggs, and works on the nest. She will also curl herself around the petiole of the nest in order to provide warmth to the eggs and larvae.

Producing the first workers

Asian hornets, like many other insects, pass through very different developmental stages before they become adults. The queen lays an egg, which she attaches to the side of the bottom of the paper cell with a strong glue. Once the larva hatches from the egg it remains attached to the egg shell, which prevents it from falling out of its downward-pointing cell. In later stages, the larva is prevented from dropping out by not wholly casting off its skin, which connects it with the egg shell. The larva will moult (cast off its skin in order to grow) four times, and is fed with balls of mashed-up insect meat, carrion and sugary liquids by workers, and grows rapidly. When the larva reaches its fifth stage, it is large enough to fill the entire cell, and it supports itself by pressing against the cell walls. At this point it loses its attachment to the wall and is able to move freely within the cell.

The larvae, especially at the last stage, regurgitate sugary saliva when requested by the adults: this saliva also contains important proteins and amino acids and is a perfect food for the queen or workers (Matsuura & Yamane 1990). After completing feeding and prior to pupation, the mature larva applies thin silken lines around the walls of the cell, and spins a thicker cocoon cap over the opening. It dumps its gut contents into the cell as it spins its cocoon and moults a final time. Thus, the larval excrement is sandwiched between the new cocoon and the final shed skin, forming a small black package called a meconium, at the closed end of the cell. Contrary to what honey bees do, these meconia are not removed by workers when cleaning cells after adult emergence, and the queen will re-use the cell up to four times (Rome et al. 2015). The number of meconia found in a cell will therefore indicate how many times it has been used.

Inside its cocoon, the final-stage larva undergoes an incredible transformation, during which its whole body is rearranged into that of an adult. This is the pupal stage of development, and the pupa is at first cream-coloured like the larva, gradually becoming pigmented towards the end of its development, and eventually biting its way out of the silken cocoon cap. The newly emerged adult (sometimes referred to as 'teneral adult' or a 'callow') is paler than older adults, has a softer skin and, in yellow hornets (*Vespa simillima*), spends the first 1 or 2 days head-down in an empty brood cell, resting and being fed by older workers while its cuticle (skin) hardens (Martin 2017). Orientation flights are taken at 1-4 days post emergence, once wings have hardened.

Above: *eggs in cells (Photo by Judy Collins)*
Below: *larva showing 'tail' made from previously shed skin*

The development of the hornet from egg to adult takes around 50 days for the first cohort of workers (Thiéry et al. 2014), but as the nest grows, and the workers are able to control and maintain nest temperature, this time will reduce. Dong & Wang (1989) found the different developmental stages in another subspecies, *Vespa velutina aurea,* to be 9-15 days for the egg, 10-18 days for the larval stage, and 15-20 days for the pupal stage in a laboratory caged primary nest (averages: eggs 13 days, larvae 15.8 days, pupae 19.3 days — overall 48.1 days for egg to adult emergence).

Early dangers

The primary nest stage in the life cycle of Asian hornets is fraught with danger. The nest must be constructed late enough in the year for temperatures to be warm enough for development of offspring. The nest itself must be protected from the elements and small mammals, yet the queen must leave the nest to forage. The queen is the sole provider, at first: if she dies the whole colony will quickly die from starvation. Around Tours, in France, single queens founded 12 colonies, but out of these only three survived (Darrouzet et al. 2014).

One danger to be overcome by a foundress is that of nest usurpation. Information from other *Vespa* species show that if an embryo nest is destroyed, it may be that the foundress doesn't try to build a new one but instead tries to find another, and fights the resident foundress for possession. Such fights often cause injuries and sometimes death, and dead foundresses can be found beneath embryo nests (14 dead foundresses were recorded under an active *Vespa affinis* embryo nest, Martin 2017). Even the winning foundress may be injured sufficiently that the embryo nest fails (Martin 2017). Another possibility is that later emerging foundresses, while searching for a suitable nest site, come across an embryo nest and gamble on winning the fight and taking over the nest. In the European hornet, queens emerge over a relatively long period — a 'bet-hedging' strategy because some of them will wake up to better conditions than others (Gourbière & Menu 2009). It is not known whether the same is true for *Vespa velutina*. Foundresses appear and are seen foraging over a longer period than might be expected if they all emerged around the same time, but on the other hand they will be flying for quite a while, setting up an embryo nest, raising the first cohort of workers and then flying for another 2 or 3 weeks once they have emerged — in total this adds up to perhaps 12 weeks.

If the queen dies at the primary nest stage or later, but there are workers present, some of the workers may lay unfertilized (haploid) eggs that will develop into males. Early males are sometimes found (Monceau et al. 2013a), but it is not known whether these are

Very small primary nest with the first workers. Photo by John de Carteret

produced by queens or workers. It has been observed that workers are hostile towards early males, attacking them and seeming to drive them away from the nest (Monceau et al. 2013b). Because workers are unable to lay eggs that will turn into workers or queens (they can only produce males), and males do not help with foraging, feeding, defence or nest building, the colony is doomed if the queen dies before the sexual stages of the colony are produced.

The primary nest grows

As soon as the first workers emerge, the whole process of nest building and brood rearing speeds up, reducing the development time of the hornets towards the end of the season (the time reduces to around 29 days in *Vespa simillima*, Martin 2017).
This is partly due to more workers constructing cells and feeding larvae, and partly due to thermoregulation in the nest. Like other bees and wasps, Asian hornets can produce heat to warm the nest by using their flight muscles. By the time a yellow hornet (*Vespa simillima*) nest is producing sexuals, the temperature in the nest is being kept between 25 and 30 °C (Martin 1990). A warm nest not only speeds up development of larvae, it also means that individual hornets do not have to warm themselves up before they are ready for action, whether that is foraging at cooler temperatures during the day (unlike the European hornet, *Vespa crabro*, Asian hornets do not fly at night) or defending the nest at any time of day or night. Conversely, when temperatures are too high on hot summer days, the workers lower the temperature of the nest through evaporative cooling. To do this, they regurgitate water onto the walls of the nest (sometimes soaking the walls), and evaporate the water through fanning with their wings (Perrard et al. 2009). This ability to regulate the temperature of the nest is one of the factors that make them such a flexible and formidable invasive species.

A primary nest in a building void. Photo by John de Carteret

The queen will still leave the nest to forage once the first workers have emerged — behaviour that carries on for 2 or 3 weeks — after that she will stay in the nest (Martin 2017), except for moving to the secondary nest if one is built. The workers take on all nest duties, except for laying eggs: they forage for protein and sweet liquids, and for water and nesting material; they construct the nest, which is ongoing until late September (France: Monceau et al. 2017); they guard the nest, adjust its temperature, clean the nest and one another, and share food.

Monceau et al. (2013b) found that older Asian hornets became the first line of defence for the colony, being the first to react and come to the door when a caged colony was threatened (by a pair of tweezers!). Another study on a caged nest (Perrard et al. 2009) found that the majority of activities consisted of foraging (47%), nest construction (16%) and exploration (13%), while the remainder involved nest guarding, cleaning, ventilation, grooming and nestmate interactions. No outside activity was seen when the temperature dropped below 10 °C. Monceau et al. (2017) found that nest patrolling and nest building were pretty constant during the day, with slightly more during the middle of the day, but that much more effort was put into foraging. All activities increased during the summer and peaked in early autumn.

Although Asian hornets do have castes (the workers are female, but perform different duties over their lifetime compared with the female queens) they do not seem to have the high level of age-related specialisation of tasks equivalent to that found in honey bee colonies (temporal polyethism). However, Asian hornets are comparatively understudied so far; there are many details of their life histories that are as yet unknown.

The nest continues to grow, with a combination of fresh material and recycled material from the inside of the nest added constantly to the outside, and material removed from the inside to make more comb. If the space that the primary nest occupies is big enough, and the location is deemed secure enough, the nest simply expands right through the season, and the initial embryo nest built by the foundress becomes embedded in the larger secondary nest (Rome et al. 2015). However, the majority of primary nests are abandoned, and a new secondary nest is built. Dissection of 69 nests in France, and the INPN database in France, suggest that approximately 70% of colonies relocate in the summer (Rome et al. 2015). Relocated nests can be distinguished from primary nests that grew into secondary nests but didn't relocate because the first (top) comb has a very regular structure in relocated nests; otherwise the top comb becomes embedded in the nest structure (Rome et al. 2015). Primary nests have their entrances at the bottom.

Relocation

In *Vespa crabro* and *Vespa simillima* in Japan, once the decision to move has been made, the workers stop their normal duties and start scouting to find a suitable location for the secondary nest. Good sites (usually within 10 m, but up to 180 m) are marked by scouts settling there, and there is some back-and-forth with workers and scouts visiting possible sites and the primary nest until some sort of consensus is arrived at and the most favoured site wins out. After 3 or 4 days of this process, the queen leaves the old nest and searches for this new nest site. When she finds it she does not return to the primary nest, and her arrival triggers the workers into constructing the secondary nest. Workers left behind in the primary nest also have to find the new site for themselves, but once they have, there is traffic between the two nests for up to a month (presumably longer in *Vespa velutina*) while all the eggs, larvae and pupae develop into workers and find their way to the new nest. Interestingly, if the new nest is taken away at night during this relocation period, workers from the primary nest will build a new nest at the site of the removed nest (Matsuura & Yamane, 1990). It seems that *Vespa crabro gribodoi* (the British subspecies) sometimes relocate their nests, starting in small cavities or small mammal burrows and building secondary nests in larger cavities (Pawlyszyn 1992). However, they also build similar embryo nests to *Vespa velutina* in sheltered locations such as sheds and barns: make sure you know which species you are dealing with.

The secondary nest

Secondary nests start as small balls that expand throughout the season, often becoming more elongated into an egg shape when they get big. Unlike in primary nests, or the nests of European hornets, the entrance is on the side of the structure in a secondary nest: it is usually circular and around 2.5 cm in diameter (26 mm, Bob Hogge, pers. comm.). Because the nest is in a constant state of construction, the outside of the nest has very distinctive structures that look like hanger openings. It is tempting to see these as entrances into the nest, but in fact they are half-finished 'bubbles' or pockets, which the hornets add to the outside of the nest walls to expand the nest. When complete, these trap air and insulate the nest. The top of the nest is made up of a dense arrangement of lots of small paper pockets, resulting in a thick, strong roof, while the rest of the nest has lighter walls made from fewer paper cells. Still, the whole nest is pretty strong — more like cardboard than paper. This material does not seem particularly waterproof (indeed, the evaporative cooling they employ relies on the water sinking in to the paper), but the exposed sites they choose in the tops of trees attest to the strength of the

Bob Hogge holds a secondary nest, which has engulfed small oak branches and twigs. This one was around 45 cm in diameter. The entrance hole can be seen in the centre. Photo by Gerry Stuart

nests, even when wet. The complex wall structure must be enough to keep the nest rigid enough until it dries out in the wind.

The first horizontal comb is built on a single petiole and subsequent combs are built below: the youngest is at the bottom. These further combs are attached to the comb above by multiple paper petioles; extremely tough pillars that are reinforced with old silk cappings from used brood cells (Martin 2017).

As further combs are built below the first, additional struts are added between the top comb and the roof to carry the extra weight of combs full of brood. The brood is arranged in concentric circles, with pupal stages covered by their distinctive domed silk caps of varying heights.

Mature, late-season secondary nests are usually very big and distinctive, once you spot them. One particularly large nest found in Aquitaine, France was 1 m tall and 80 cm across (Rome et al. 2015), and a typical secondary nest from a tall tree and dissected in Jersey in 2018 was approximately 48 cm tall by 42 cm wide. Interestingly, the nests found in the UK in 2018 were all fairly small: 20-25 cm in diameter (Nigel Semmence, pers. comm.).

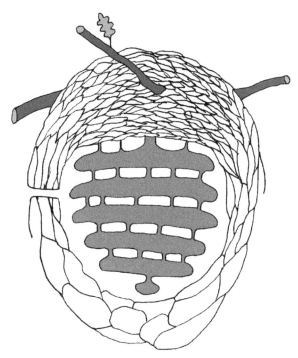

Cross-section through a secondary nest. Note the more compact paper pockets at the top, entrance hole on left-hand side and unfinished pockets.

Although secondary nests are stunning structures to behold, they are still very difficult to pick out amongst foliage, especially high up in trees.

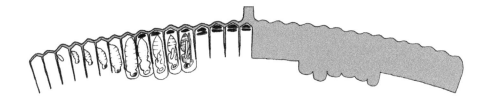

Cross-section through comb showing developmental stages from egg to larva, and finally pupa. The black objects in the bottom of the cells are meconia. Notice that the fifth-stage larva is free from its tether.

Secondary nest cut away to reveal combs inside. Photo by Judy Collins

Nests can be found in forests and agricultural areas, but in France 48.5% were discovered in urbanised habitats (Rome et al. 2015); whether this is due to increased likelihood of discovery by more observers, or whether they thrive in urban areas is not known, although Martin (2017) cites research that he was part of (Choi et al. 2012), which looked at the dominance of Asian hornets among five species of hornet in Korea: in this case Asian hornets became more and more dominant as the habitat became more urbanised, with Asian hornets being almost the only type of hornet found in cities. It is not known, however, whether Asian hornets prefer urban environments in Korea, or whether they are taking up residence in cities to avoid competition with other species of hornet.

A study by Monceau & Thiéry (2017) looked at distribution of nests in the small coastal town of Andernos-les-Bains on a sheltered bay west of Bordeaux, France. The study looked at data over 8 years (2007 to 2014), with a total of 528 nests. They found that in 6 of the years, the nests were randomly distributed within the town, suggesting that competition for nesting sites and/or food supplies did not exist. When the data was combined for all 8 years, it showed an aggregation of nests close to the seafront where there was an oyster farm and sport fishing activities; the authors suggest that these provided a food source that favoured colony development. By the end of the study, the density of nests had reached 12.26 nests per square kilometre. The same data were used by another group to try to estimate the carrying capacity (i.e. the number of nests that an area can support, long term), and arrived at an average density of 10.64

nests per square kilometer if nests were not detected and destroyed and 8.07 nests per square kilometer if they were. Interestingly, they also calculated that the relatively high detection rate for nests was only detecting about half of the nests actually present, and only 37.4% of active nests were discovered (Franklin et al. 2017).

Franklin et al. (2017) also looked at the structures that nests were built on. The data showed that preferences are very different for primary and secondary nests. Out of 39 primary nests, roughly 77% were located in or on man-made structures, and the remaining 23% on natural structures. For secondary nests, these proportions reversed, so that (out of 225 secondary nests) roughly 78% were on natural structures (of which roughly 94% were trees and 6% were other natural structures, like bushes and hedges) and 22% were on man-made structures. Three nests (1.5%) were found underground. The trees that were most popular were oak (64.3%), pine (15.9%), plane (4%), and poplar (2.4%): other species were represented at smaller percentages.

In Jersey in 2018, out of 36 secondary nests, 7 were found in sycamores, 6 in building voids (like inaccessible roof spaces) or in a barn, 4 in oak trees, 3 in hedges, 2 in bramble patches, 1 on the outside of a building and the rest in assorted trees (copper beech, ash, pine, lime, apple, black poplar, ornamental cherry, cyprus, holly). As far as height went, of those in trees with heights estimated, around 20 were 10 m or above (15 were at or above 20 m), and another 9 were below this height and therefore potentially more dangerous for humans.

Close-up of comb with large larvae and the silk-capped cells which contain pupae.

A single comb showing concentric circles of larvae and pupae. The notch is due to the comb being built around a branch.

For the Morbihan, France 2017 nest data, almost 75% were found below 10 m (over 50% were at 0-5 m). Although this is probably a combination of data for primary and secondary nests, they recorded 700 primaries and over 3000 secondaries, so the data has a much larger contribution from secondary nests. From the accompanying notes, it sounds as if there is a continuing trend for lower nests, and for more nest sites on man-made structures (42% trees, shrubs and hedges, 58% on man-made).

Whether this reflects an overall change in choice of nest site, or whether it is due to the most likely observations being in urban areas, where natural substrates are limited, is unknown (FDGDON-Morbihan 2018).

Where Asian hornets have been established for a few years in France, people have discovered 'hotspots' or favourite nesting sites where

Hotspot. This tree in France had four nests in it, which were all active during the same season. Photo by Felix Gil

Asian hornets will build nests in the same tree year after year, or even in the same year.

By adding up the estimated numbers of eggs, larvae and meconia for each comb, 21 mature nests (meaning they had already started to produce sexual individuals) taken down in France in October were found to have an average of 6.3 combs, around 5000 cells, and were estimated to have produced an average of around 5300 individuals up to that point. However, the sizes of the nests did vary widely, resulting in a wide range in the estimates of individuals (minimum estimated as 82, and the maximum 13,340, Rome et al. 2015). It is not known how long *Vespa velutina* workers live for: wild Asian hornets caught for flight mill experiments lived for several weeks, but there was a significant decline in their flight ability after 2 weeks (obviously their age at capture was unknown, Sauvard et al. 2018). Poidatz et al. (2018a) used RFID (radio frequency identification) to investigate activity and foraging ranges of Asian hornets, and individuals were detected for up to 26 days, suggesting that they live longer than this (they would not start to forage for a while). Dong & Wang (1989) give 24-142 days for life spans of workers and 4-60 days for life spans of males, all from a captive colony of *Vespa velutina auraria*.

Foraging

Hornets forage during the day, and the distances they travel have been under intense investigation by scientists, as foraging has such a bearing on honey bee predation and control methods, including finding the nest. The evidence comes from several sources: plotting observed individuals on a map around a nest (only possible when you know you have one nest and they come from it), marking individuals and timing them to the nest (Jersey), fitting individuals from a captive nest with tiny RFID tags (the hornet crawls through a tunnel when leaving or entering the nest, which reads the tag, thus they can be timed, or released elsewhere and timed home), or using a flight mill to find out how fast they can fly, and for how long (a flight mill consists of a small stand with a freely rotating wire arm to which the insect is tethered).

In the case of Asian hornets, a consensus is forming that the majority of foraging occurs under 700 m from a secondary nest (NBU, Sauvard et al. 2018, Poidatz et al. 2018a), and perhaps 350 m from a primary nest (early summer, Sauvard 2018). However, there are also long-range foragers that will go up to 5 km (Poidatz et al. 2018a).

This same RFID study (in France: they used a very young nest in early June) also found that only approximately 5% could get home if released 5 km away, and their return did not seem to be influenced by the direction of the release point from the nest (NW, NE, SW, SE), perhaps showing that they use visual and scent landmarks rather than sun and geomagnetic cues. They also found that the ones to make it back to the nest tended to be smaller and therefore were probably older and more experienced. Around 50% or fewer were able to get back to the nest when released at or above 1 km away.

When they looked at times spent foraging, they found that the vast majority (95%) of trips from the nest lasted less than an hour, and that there was a tiny amount of nocturnal activity (2%: but this may have been a walk rather than a flight). Finally, there were hornets that went for particularly long trips, perhaps scouting, and some which did not come back for days — did they rest for the night in vegetation? And what of the hornets that did not return during the experiment: did they die or were they accepted by another nest? [Perrard et al. (2009) found that some Asian hornets introduced from other nests to a caged nest were accepted].

Sauvard et al.'s (2018) study of tethered hornets gave an average flight speed of 1.56 ± 0.29 m per second, but the authors note that in field observations, *Vespa velutina* is more likely to fly at speeds similar to free-flying *Vespa crabro*: around 6 m per second. Empirical observations from Jersey, timing marked insects, and plotting the release point and nest on a map, have resulted in a rough rule of thumb of a minute's absence from the bait equating to the nest being around 100 m away, so return flight time = 200 m / 60 s = 3.3 m per second (with no 'unloading time' considered). Peter Kennedy was able to measure flight speeds of radio-tagged Asian hornets, and recorded an average flight speed of 2.9 m per second (± 0.6 m/s, n=9), and untagged hornets fly as fast, if not faster, than tagged hornets (Peter Kennedy, pers. comm.)

Foraging activity is throughout the day, with examples given of 7 am-8 pm (Poidatz et al. 2018a); main activity between 8:30 am and 8:30 pm (Perrard et al. 2009); 8:00 am to 8:00 pm, with a small peak in all activities, including foraging, between 1:00 pm and 2:00 pm (Monceau et al. 2017). The early start is perhaps related to the possibility of collecting dew that is present during the early morning, as well as the increased demand from larvae after a night without food. (Matsuura & Yamane 1990). Jersey trackers observed workers foraging in strong winds, drizzle and rain.

Food for adults

In order to sustain the rapid growth of the colony from the summer to the autumn, the workers forage for sugar-rich liquids (carbohydrates) for themselves and protein for the larvae. They are able to exchange food with other adults in the nest (a process known as trophallaxis) and, as previously mentioned, obtain food in the form of special saliva from later-stage larvae. Indeed, in some ways the larvae act as food storage for the whole colony (Matsuura & Yamane 1990). Hornet larvae are primarily fed with chewed prey pellets, and signal to the adults when they are hungry by gyrating in their cells, scraping the sides of the cell walls to produce a noticeable rustling/rasping sound. The adults themselves cannot digest solid food (it can't get through their narrow waist), so rely on larval regurgitations, tree sap (when gathering nest material, fresh plant material is often moist with sap, and trees exude sap when broken or wounded by birds, insects, mammals or fungi), floral nectar (this may be limited to those flowers where the nectaries are easily accessed due to their short tongues: *Camellia sasanqua*, banana flowers and ivy were the most obvious in Jersey), the juice from meat (insect prey or vertebrate carcasses), ripe fruit and man-made foodstuffs like sugary pop drinks or anything else liquid and sweet. A list of flowers used by Asian hornets in Japan can be found in Ueno (2015).

Asian hornet investigating a banana flower on Jersey. Photo by Gerry Stuart

Food for larvae

When the nest is still small (in early spring) there is just the queen, or later a few workers feeding larvae, resulting in small adults. As the season progresses, more workers forage, so the larvae, although more numerous, get larger amounts of food, resulting in bigger individuals. The size of workers increases and their mean weight doubles (188.8–

386.4 mg) from June to November (there is a lot of variability in these weights: Rome et al. 2015).

There are plenty of observations of Asian hornets taking meat from wild carcasses and from butchers and fishmongers. In cities there are many opportunities for scavenging man-made sources of discarded protein. When it comes to hunting, Asian hornets are formidable, and this is what causes concern for both beekeepers and conservationists. Villemant et al. (2011b) found that *Vespa velutina* consumes a great variety of insects, with an overall preference for social hymenoptera (honey bees 37%, social wasps 18%) and flies (34%, mainly hover-flies, blow-flies and house-flies and their close relatives). Predation on honey bees was worse in an urban environment where there was less overall choice of prey.

Two Asian hornets gathering flesh from a small mammal carcass. Photo by Chris Luck

Prey may be caught on the wing or on a surface, and is grabbed and bitten. After capturing its prey, the hornet worker either takes the whole insect back to the nest or processes it to some extent, from removing some parts (head, wings, legs, perhaps) to making a flesh pellet exclusively from the protein-rich thorax (which contains the big flight muscles). To make a 'meat ball', the hornet retreats to a leaf or twig and hangs by

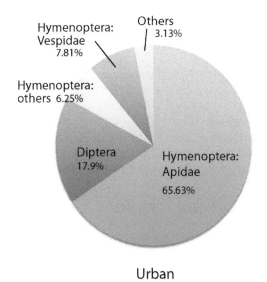

Urban

one or two back legs, using its other legs to manipulate the body of its prey. It removes the legs, wings, abdomen, head and sometimes the thorax cuticle, a process taking around four and a half minutes for a honey bee caught by a hornet from a captive nest (Perrard et al. 2009).

Two studies have caught and analysed pellets brought back to the nest by returning foragers: Villemant et al. (2011b) collected more than 2500 prey balls between 2007 and 2009 in urban, agricultural and forested areas in France, with different results (see pie charts below). Perrard et al. (2009) collected loads from hornets returning to a wild urban nest during July 2007. They collected 235 loads, of which 71.8% were prey and 27.2% were pulp for nest building. Of the prey, 84.8% were honey bees, 11.7% were other insects (mostly halictids and Brachycera) and 3.5% were vertebrate flesh, of which a third contained feather remains).

Van der Vecht (1957) notes that in Indo-Malayan and Papuan areas, *Vespa velutina* were seen hunting and feeding on honey bees, stable flies (*Stomoxys calcitrans*), a blowfly (*Chrysomyia megacephala*); in Java they were seen to feed on tachinid flies and take a spider from a web.

Below, this page and opposite*: pie charts after Villemant et al. (2011b). Note that 'Hymenoptera: Apidae' includes wild bees as well as honey bees, but the wild bees (mainly bumblebees) are very few compared to honey bees (Claire Villemant, pers. comm.)*

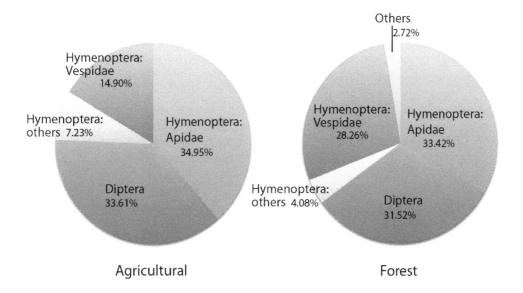

Honey bee predation

When such a high percentage of Asian hornet prey can be honey bees, you can see why UK beekeepers are so alarmed at the idea of a *Vespa velutina* invasion into the British Isles.

Honey bees are perfect for Asian hornets — they're big, with plenty of meat on them, and humans keep them in these boxes in open areas where they are easy to attack. Not only that, but usually there are several boxes in one place! Tens to hundreds of thousands of prey items in one place — this is fairly easy, fast, food.

Asian hornets are 'central place' foragers, meaning that they need to keep returning to the nest to feed larvae, and in a world of patchy food supplies, a hive or apiary is the most efficient way of providing the nest with food. It is not known whether nestmates can communicate the position of food sources to one another, like bees: perhaps workers bringing in good food are followed to the source. Some other *Vespa* species (like the Asian **giant** hornet, *Vespa mandarinia*) actually mark sources of food, such as hives, to signal to their nestmates, and will raid a hive en mass: fortunately *Vespa velutina* does not show this type of aggressive raiding nature.

There is some evidence that individual Asian hornets might stick to food types, at least temporarily. The NBU, when tracking marked individuals in the UK, used both fish fillets and wasp attractant. They found that the same individuals would consistently return to either protein or sugars, perhaps indicating that a worthwhile patch of food should be exploited until it starts to dry up. Within a food-type (in this case honey bees), Thiéry et al. (2014) found that more than half of attacking Asian hornets returned to the hive they were caught at to carry on predating, but Monceau et al. (2014b) found most of their hornets (88%) were recaptured in front of different hives.

Couto et al. (2014) found that Asian hornets captured in front of hives were especially attracted to the odours of pollen, honey, and brood, even more than adult honey bees, and meat and fish were ignored. Hornets were also attracted to the odour of geraniol, which is a component of the Nasanov pheromone in honey bees. Geraniol is a 'come hither' (aggregation) pheromone, gathering bees to a swarm or the hive entrance, and may indicate to hornets the presence of large numbers of honey bees, so it is thought that *Vespa velutina* mainly finds its prey through scent.

Asian hornet biology

Once a hive has been located, Asian hornets will hover in front of it and attack mainly returning foragers, a behaviour known as 'hawking'. Returning foragers may be preferred because they can be 40% heavier when they return with nectar and pollen loads, which will affect their speed and agility; also, being the oldest workers, wing damage may also affect their flight (Monceau et al. 2013c). However, hornets will also attack bees on the ground and on the hive, especially if they get isolated from the crowd.

Predation pressure increases over the season, from the first Asian hornets spotted in an apiary: 40 days after these are seen, the period of intense predation begins (July, Monceau et al. 2013c) and extends to the end of October or November (French data), with a maximum in mid September and a drastic decline at the end of the season. During a day, Asian hornet numbers in front of hives are fairly constant, but they are more efficient in catching honey bees in the middle of the day (1-2 pm); meanwhile, honey bee flying activity reduces during the day. Temperature and humidity did not seem to affect numbers, but wind decreased the numbers trapped at an apiary, suggesting that they are less likely to forage on windy days, although wind did not seem to deter Jersey Asian hornets. It is important to remember that Asian hornets are not specialist honey bee predators; they are generalists that take advantage of colonies of honey bees. Monceau et al. (2013c) found that the greatest number of honey bees were captured with nine hornets hawking in front of a hive: more hornets did not catch more honey bees. The authors wonder whether the

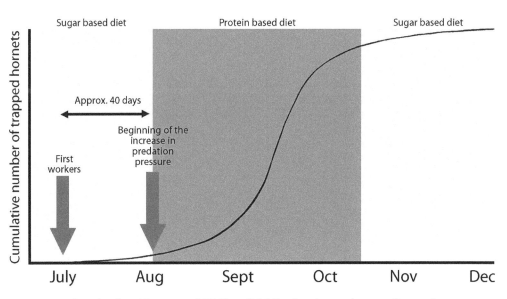

Graph after Monceau & Thiéry (2016): the three phases of trapping

Asian hornet dismembering a honey bee. Photo by Peter Kennedy

extra hornets may be acting as guards, perhaps defending this food patch from other Asian hornets, or whether competition was occurring between hornets.

In Kashmir, up to 25 *Vespa velutina* were seen hawking in front of a hive (Shah & Shah 1991), while a maximum of 20 were observed by Monceau et al. (2014b) in France. Shah & Shah observed an average of 24 *Apis mellifera* honey bees (European honey bees) being caught in a 30-min period, in an apiary with three hives. By extrapolation, that would lead to 576 honey bees being caught over a 12-hour foraging day, if the capture rate remained constant. If this observation was made in the middle of the day, then the daily catch would be somewhat lower. It was unknown in this study how many Asian hornet nests were contributing to attacks on the apiary, but the overall effect would seem to be about as bad as it can get. More data on catch-rates of honey bees by Asian hornets is needed.

Scientists in France have looked at predation from the individual Asian hornet point of view. At one apiary in France, with at least five Asian hornet nests within a kilometre from the site, it was estimated by capture-mark-recapture methods that at least 350 individual hornets were visiting the apiary per day in early August, and made a couple of visits in a half-day. Over the 9-day experiment, there was a turnover in individuals, with only four individuals visiting on all 9 consecutive days. Two thirds of those marked first returned at least once: overall at least half were captured more than once. Interestingly, the hornets did not prefer the largest or the smallest colonies: one hive in particular was preferred by the hornets, and the authors wonder whether this colony was least aggressive. Finally, as seen in Jersey when timing hornet visits to bait stations, subsequent visits by the same individual became faster, suggesting that they became more efficient in finding their way (Monceau et al. 2014b).

Late in the year, when the temperature is low, Asian hornets may enter a very weak honey bee colony to steal pollen and honey, scavenge on recently dead bees and attack isolated ones (Shah & Shah 1991). Or indeed, if the colony is sufficiently weakened earlier in the year, such that hive activity and guard behaviour is barely seen, Asian hornets will enter the hive to steal honey, pollen, larvae and pupae (Arca et al. 2014).

Honey bee defence

In the Asian hornet's native range, the honey bees that they co-evolved with are a different species (*Apis cerana*) to those of western Europe (*Apis mellifera*), and it turns out that this has a big impact on how well European honey bees can defend themselves.

Apis cerana have evolved several strategies to defend themselves against *Vespa velutina*. They use an avoidance tactic: normally, both *Apis mellifera* and *Apis cerana* approach the hive slowly and alight near the entrance, sometimes making 'sashaying' movements (slipping from side to side) as they come in. However, when *Apis cerana* is under attack from *Vespa velutina*, they speed up as they come in to land, making a straight dash through the most dangerous piece of air-space twice as fast as *Apis mellifera*, which come in slowly, making an easier target. Also, on take-off, *Apis cerana* are more likely to fly straight off and *Apis mellifera* are more likely to circle before heading off (Shah & Shah 1991).

Apis cerana recruit three times as many guard bees to protect the hive as *Apis mellifera*, and have another couple of tricks to keep themselves safe. The first of these is 'shimmering' where between 10 and 200 workers gather around the hive entrance and shake their bodies from side to side. This sends an 'I have seen you' message to the hornets, and keeps them about 15 to 30 cm away. The bees shimmer more violently if a hornet approaches more closely (Shah & Shah 1991). *Apis mellifera* have not been observed to do this. Both species do make 'bee carpets' essentially a co-ordinated blanket of tightly packed bees around the entrance or on the landing board of the hive – a prerequisite for shimmering – and both species can produce a hissing sound (Arca et al. 2014). In *Apis cerana*, if a hornet makes contact with a bee carpet, it can be pulled in by a couple of guard bees and is rapidly attacked by 30-60 bees, grabbing, biting, holding and stinging it. Initially a loose ball of bees is formed around the hornet, but gradually some bees retreat, leaving around 25 hard-core defenders that form a tight bee ball around the hornet for 30 min to over an hour, and the hornet dies from overheating, or possibly overheating combined with increased CO_2 and humidity. Another mechanism of hornet death in a bee ball has been discovered in the southern European hornet *Vespa orientalis*: in this case the bees (a subspecies of *Apis mellifera*: *Apis mellifera cypria*) asphyxiate the hornet by restricting the movement of its abdomen, which cuts down its ability to breathe [hornets and wasps 'pant' by expanding and contracting their abdomens, pulling air in and out through a series of holes (spiracles) along their sides]. Stinging may also be a factor in bee balls. Both *Apis mellifera* and *Apis cerana* will also build propolis walls to narrow and block off entrances during attacks. With this possible repertoire of behaviours, the European bee does not do very well. They can form a bee carpet and hiss, but they rarely form a killing ball and do not shimmer. All in all, their defence is pretty poor. The result is that when given the choice between attacking *Apis cerana* or *Apis mellifera*, Asian hornets prefer attacking *Apis*

*Japanese honeybees (*Apis cerana japonica*) forming a "bee ball" in which two hornets (*Vespa simillima xanthoptera*) are engulfed and being heated. Yokohama, Kanagawa prefecture, Honshu Island, Japan. Photo by Takahashi (Own work, CC BY-SA 3.0)*

mellifera, and will be three times more frequently found at *Apis mellifera* hives, taking *Apis mellifera* eight times more frequently than *Apis cerana* and being three times more successful at catching *Apis mellifera* than *Apis cerana* (Tan et al. 2014). In Kashmir, Shah & Shah (1991) observed on average one (range 0-2) *Apis cerana* caught in half an hour compared with an average 24 (range 12-44) *Apis mellifera*.

Asian hornets hunting at a hive in France. Two are walking on the hive body, perhaps looking for an isolated bee; three others hawk in front of the hive. Photo by Felix Gil

Being under attack by Asian hornets firstly results in a lot of stress for the whole colony, and attacks are at a maximum in the critical pre-wintering season for the bees when they need to get enough stores in to see them through the winter. There are two paths that lead to overwintering collapse. First, there is the reduction of foragers by predation, which can lead to too small a population of bees to get through the winter, and second, lots of hornets cause 'foraging paralysis' where workers refuse to go out foraging because it is too scary. This results in a greater population trapped in the hive and eating the stores, leading to a food shortage in the winter. Collapse is rarely seen in full-sized colonies during the hornet attacks: it is these other things that lead to winter collapse. When foraging paralysis

Asian hornet hawking in front of a hive in France. Photo by Felix Gil

occurs, it is even easier for hornets to pick off the few individuals who do fly (Requier et al. 2018). The authors suggest merging small colonies and supplemental feeding with bee bread (pollen and honey) at low Asian hornet loads, and apiary trapping and muzzles at high Asian hornet loads, along with supplementary feeding both pre-winter and winter: something that French beekeepers have become used to doing.

In a context where honey bee colonies are already weakened from parasites, pesticides and lack of good forage, this is yet another problem for their survival.

Production of the sexual stages (fertile males and gynes)

In a French study, the number of workers reached a maximum between late October and early November, with an average of around 436 workers present: the most workers recorded was 1742 (Rome et al. 2015). In the Alresford nest in the UK, which was removed 22nd September, 122 adults were found (the nest was 25 cm in diameter).

Autumn is when the colony reproduces by creating fertile males, and females (gynes) who will become queens of colonies next spring.

It is not known what mechanism leads to the production of gynes rather than workers (they are both female), but by September the queen switches to laying eggs that become males and then lays eggs that will become gynes. On average, three times more males than gynes are produced from mid-September to the end of November (on average 350 gynes vs 900 males, Villemant et al. 2011b). Chauzat & Martin (2009) mention a positive correlation between the nest size and the number of queens raised; and, certainly, the longer favourable conditions continue in the autumn, the more gynes will reach maturity. Hence the race to find and destroy nests in the autumn before they have a chance to produce these 'sexuals'.

In an observational study of 77 nests in France, Rome et al. (2015) reported seeing males as early as July in nests with laying workers or perhaps from queens that had not been well fertilised. However, in nests with 'normal' development, males emerged in early September, about 15 days before reproductive females. During the male's first 10 days of life, spermatogenesis (the migration of sperm from testes to seminal vesicles) is completed, and the males are then sexually mature, leaving the nest and starting to forage for themselves (Poidatz et al. 2017). Newly emerged males and gynes will spend their first 1-2 weeks based in the nest, feeding on larval and worker regurgitations; the gynes, especially, need to build up their fat reserves to see them through their hibernation.

In the French study, both males and gynes reached maximum numbers during the first half of November and most had left the

Male Asian hornet showing underside with distinctive pair of yellow spots near tip of tail. Photo by John de Carteret

nest by the end of November. In December the last workers were still present and feeding the sexual brood, although due to prey scarcity, the brood was unlikely to reach maturity (Rome et al. 2015).
The same authors found that the last nest still containing a mother queen in their study was as late as 27th October.

The reproductive strategy of hornets is based on the synchronicity between the emergence of sexually mature males and gynes (Poidatz et al. 2017) and Matsuura & Yamane (1990) report that males and queens of other Vespine wasps often leave the nest at the same time.

Mating

Genetic analysis of nests found in the UK in 2018 found that queens had mated with one or two males (Nigel Semmence pers. comm.), and in France, research has shown that gynes mate with between one and eight males (Arca et al. 2015). The details of how mating occurs have been somewhat unclear, and have been difficult to study. So far, there have only been a few observations of *Vespa velutina* mating in Europe, and these have been seen on surfaces rather than on the wing (for example on a sunny pavement, K. Monceau, pers. obs. in Monceau et al. 2014a), or on the ground adjacent to nests or food plants (Poidatz 2017, AAAFA in Poidatz 2017), where they form an 'S' shape together.

In China, Wen et al. (2017) have really moved the research on by finding two compounds that act as sex pheromones in gynes of *Vespa velutina* (although not the same subspecies as in Europe, *nigrithorax*), by using a combination of observations of 'feral' colonies and obtaining and analysing the compounds from newly emerged captive gynes. They found that virgin gynes would leave their nests on sunny days and head for open areas at the rim of a nearby forest. The gynes would repeatedly fly out into these areas to attract males, then fly back to a more sheltered location within the trees. The gyne would rub herself with sex pheromone, which would attract swarms of males, but after mating, she would no longer produce these compounds.

Once a male successfully finds a gyne and is accepted, copulation occurs. Males do not die as a result of mating, like in *Apis mellifera*, and work by Poidatz et al. (2017) suggests that each male only mates once, eventually dying around December. Although males may forage for their own sustenance during their search for mates (ivy being a well-known autumnal source of nectar and pollen for many insects, Spradbery 1973), the plants visited by males during the autumn may also represent a rendezvous site for both sexes; if so, then gynes should also occur on these plants.

Nest decline

Once the sexuals have left the nest, the numbers of workers decline, due to a combination of fewer sources of food, colder temperatures, their own life spans coming to an end and lack of replacement by new workers, until the nest is fully abandoned. Rarely, one or more gynes can be found in an empty nest; these have emerged too late to mate and often have atrophied wings. Since they are unmated, they will not be able to found a nest (Villemant et al. 2011b). The last of the larvae and pupae in the nest die of starvation as the food resources run out, and the last few workers concentrate on feeding brood destined to be sexuals; however, due to prey scarcity at that time of year, some of these will never reach maturity.

The nests are not re-used (although there has been a case of an embryo nest being built inside a secondary nest which had been stored in an out building). Over the winter they become silvery-grey with age. With winter weather many break up and fall to the ground.

Green woodpeckers, jays and tits will eat the last of the larvae at the end of the season (Mollet & de la Torre, 2006).

Hibernation of the queens

After mating, the queens spend time searching for hibernation sites. The dispersal range of queens is unknown, but it is thought that mated queens do not make long-distance dispersal flights, perhaps hibernating from several tens of meters to over 1 km in order to conserve resources for hibernation, as they don't eat after copulation (data for other *Vespa* species: Matsuura & Yamane, 1990). However, humans can accidentally transport them great distances. The ovaries of hibernating queens remain in an immature condition, and the sperm they have stored in their bodies is conserved for the colonies they will build next season. The fate of unmated gynes is unknown.

Ecology

Problems with invasive alien species

Species evolve in concert with everything around them — geography, geology, climate and other species, in a complex web. So when a species is transplanted, either on purpose or by accident, there can be impacts on what was a relatively stable network. Thus, the major worry with an invasive alien species is the likelihood that it will have a negative effect on biodiversity. In particular, invasive generalist predators can have an impact on the whole ecosystem by preying on naïve native prey, without being kept in check by natural enemies (Monceau et al. 2015). If the species in question is a top predator that is able to spread rapidly and has the versatility of a superorganism, then its effects will be dramatic, as has been the case with *Vespa velutina*.

As we have seen, *Vespa velutina* feeds on insects, some spiders and others to feed its larvae, which are carnivorous. This will have an impact on native insects directly, and then possibly beyond, on behaviours such as pollination. Carisio et al. (2018) looked at the effects of *Vespa velutina* on wild bees in the north-west of Italy (Liguria). In a piece of good news, they found similar species richness of wild bees in areas with and without *Vespa velutina*, and could work out from the seasonal appearance of different bee species which might be more vulnerable to predation from Asian hornets.

Effect on honey bees

As we have seen, honey bees are a significant part of Asian hornet predation, resulting in winter losses of colonies in those areas where the hornets have become well established. It is not yet known whether *Vespa velutina* is also capable of transferring diseases to honey bees: it seems that it can host Israeli acute paralysis virus (IAPV), which affects European honey bees in France and China (Blanchard et al. 2008, Yanez et al. 2012). Loss of colonies puts stress onto beekeepers too, (Monceau et al. 2014a), with loss of income for professional beekeepers, and some amateur beekeepers simply giving up, as was the case when varroa (*Varroa destructor*, a mite which parasitises bees) arrived in the UK. From research conducted so far it seems likely that *Vespa velutina* is going to affect native insect numbers and diversity, but it is unknown to what extent (research is underway to try to clarify this). An impact on insect-pollinated crops may be anticipated if pollinator communities are affected, but such losses have not been recorded. As well as honey bees and the pollination services they provide, the products which come from farming bees (honey, wax, pollen and propolis) are reduced when colonies are lost (Ferreira-Golpe et al. 2018).

The data from France on honey bee colony losses appear patchy, with some areas reporting greater losses than others. There is very little data available on actual colony losses, but the figure of 30% destroyed or weakened is generally referred to (Monceau et al. 2014a). Shah & Shah (1991) found that *Vespa velutina* limited colony development by persistent predation of adult bees in Kashmir. Colony losses are recorded in France (beekeepers have to register their honey bee colonies, and surveys of colony losses are well reported); the problem is that it is difficult to put losses down to any one cause when pests, diseases, lack of forage, insecticide use and especially weather can all cause colony losses. For example, in the UK the British Beekeepers' Association found overall colony losses of 13% for winter 2016/2017, but 25% for 2017/2018. This was attributed to a severe winter and very late spring in 2017/2018. Meanwhile, the ESA (Epidémiosurveillance santé animale) reported overall French losses of 30-35% for winter 2017/2018. Anecdotally, some French beekeepers are reporting 40% losses and worse but again, losses are patchy (Kevin Baughen, pers. comm.). Decante (2015) reported losses of up to 60% in colonies at the end of winter in Alpes Maritimes, France.

Effect on European hornets

On the other hand, hornets may be good at general pest control in the countryside. Indeed, *Vespa crabro*, the European hornet, was introduced into North America in the 19th century in order to control outbreaks of forest caterpillars (Villemant et al. 2006). Interestingly, *Vespa crabro* and *Vespa velutina* are both present in the part of China from where Asian hornets were accidentally transported (along with several other species of hornet). But *Vespa crabro* in Europe is not used to sharing space and resources with *Vespa velutina*, and the European hornet is under threat already in some parts of Europe (in Germany it is protected). In Jersey Asian hornets feeding on tree exudates would leave if a European hornet turned up (Bob Hogge and Peter Kennedy pers. comm.). Although European hornets do occasionally take honey bees at hives, the attacks are relatively rare and small-scale (although large-scale attacks on honey bee colonies do sometimes happen), and not something that beekeepers worry about.

Native hornets and Asian hornets are active during the same seasons (the latter are active over a longer period), eat the same things and occasionally overlap in choice of nest sites. This would suggest that there might be competition between the two species, although European hornets often re-use nests (Chinery 1986), and subtle differences in timings of different food sources may reduce such competition (Monceau et al. 2015). Not only have native hornets been observed predating more on honey bee colonies since the introduction of Asian hornets in France (Monceau et al. 2014b), but they have also been seen scavenging on dead bees left by the Asian hornets (Monceau et al. 2015). So it might even be advantageous for the native hornets to live alongside Asian hornets and such 'synergistic predation' only increases the losses for honey bees. Recent research (Cini et al. 2018) found that the two hornets' diets overlapped, no differences were found in workers' exploratory behaviour, and the natural antibacterial activity in workers of native hornets was greater than in Asian hornets.

Competition among Asian hornets

One might expect that 'pioneering' Asian hornets moving into new territory might have different population dynamics compared with the situation after a few years when there might be more competition between colonies for prey and nesting sites. When Monceau & Thiéry (2017) did their study on Andernos-les-Bains (south west France), they found that as the number of nests increased from four in 2007 to 111 in 2014, the nesting sites (apart from in one year) appeared randomly located, which would suggest that there were no territories and thus no competition between the colonies. This was corroborated by Franklin et al. (2017), who, using the same data, suggested that the high densities of nests involved pointed to a lack of competition for resources. In fact, they came up with some astonishing density levels of around eight nests per square kilometre with continued destruction of nests as an eventual carrying capacity of the area, but when this was adjusted to the purely urban part of the area, a carrying capacity of around 23 nests per square kilometre (with ongoing nest destruction) could be expected!

Effects on humans

As we have seen, Asian hornets affect honey bees and therefore beekeepers, and possibly horticulturists who rely on honey bees for crop pollination. Fruit growers in France have been affected by Asian hornets feeding on ripe fruit and pickers getting stung: there is a cost to 'red fruit' farmers, and to others who work with trees. If they have an effect on other insects (lowering their numbers), then this will have knock-on effects that could be wide-ranging, from pollination to songbird numbers. These potential impacts on nature could reduce the resilience of ecosystems that are already under many stresses.

Another impact of Asian hornets on humans, apart from economic and ecological, is the harm that can occur if someone gets too close to a nest and gets badly stung. As we have seen, Asian hornets thrive in urban settings and can reach high densities; when this is the case, more people are going to accidentally disturb nests while going about their normal activities (cutting hedges, fixing roofs, reading utility meters, etc.). People will have to learn to be more careful in these activities. Apart from actual harm, there is also fear of these insects, which can be fuelled by the media, leading to pointless worry and the killing of many other insects through misidentification.

Asian Hornet Stings

Wasps, hornets and honey bees can all deliver painful stings, and unfortunately for some people who are allergic, a single sting can lead to a severe reaction and even death. The Asian hornet, being quite a large insect, can deliver a deep and painful sting and can inject more venom than a smaller wasp.

According to a French pharmacologist who specialises in toxins at the anti-poisoning centre in Angers, Gaël Le Roux, the venom of the Asian hornet is a little less toxic than that of its European counterpart (*Vespa crabro*), but can still trigger a very serious allergic shock (Ouest France, 2015). Scientists in China have analysed the components in *Vespa velutina* venom and found toxins that affect the integrity of blood and how it clots, and neurotoxins (which affect the nervous system) that are responsible for the toxic reactions and allergic effects (Liu et al. 2015).

Honey bees have barbed stings, which get trapped in the elastic skin of mammals so effectively that when they pull away the stinging organ is ripped from their bodies and they go on to die (although not without carrying on the fight for a while first!). Wasps and hornets have smoother stings, which allow them to sting multiple times without getting them caught in the skin of their victim.

Asian hornet stings. Photo by John de Carteret

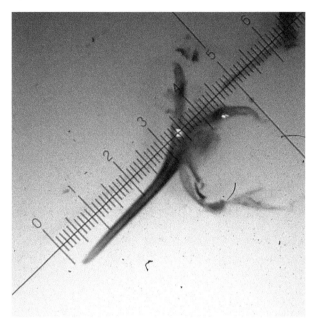

Photo through a microscope of an Asian hornet sting dissected from a worker. Scale is in millimetres. Photo by Bob Hogge

The sting of a worker Asian hornet is around 3 mm long (Bob Hogge, pers. comm. — see photo) and the specialist suits that are used by professionals who destroy and remove nests are 5 mm thick or more. This is achieved by padding or special materials that have an intrinsic open architecture.

The implications for public health and the worries about honey bee losses are at the forefront of most grassroots Asian hornet organisations in France. Already several people have died in France due to Asian hornet stings, but according to a study by de Haro et al. (2010), there has been no increase in the number of hymenopteran stings in areas of France where *Vespa velutina* has become established.

There seem to be two dangerous situations when it comes to Asian hornet stings: the first occurs when the victim is predisposed in some way to a serious outcome, such as having an allergy to stings, or a history of heart problems, in which case even a single sting can be serious, as with other species of hornet or wasp. The second situation is when the victim receives multiple stings, which only occurs when they get too close to a nest. If you accidentally get within around 5 m of a nest (de Haro et al. 2010), a full-scale attack may be triggered with many or all the hornets in the nest being called on to see you off.

An important point here is that, like honey bees, Asian hornet venom acts as an alarm pheromone, which attracts more workers and causes them to attack (Cheng et al. 2017). You don't even need to be stung to trigger this response: if the hornet is alarmed enough it can exude some venom from its sting, and the volatile components released will stimulate others to attack.

Martin (1995, 2017) says that in many parts of Asia *Vespa velutina* has a particularly aggressive nature around mature nests, and is greatly feared. In Indonesia, Vietnam and China the species is thought of as aggressive (de Haro et al. 2010). Indeed, Liu et al. (2015) describe it as 'powerful and deadly...the most aggressive and fearful species in China.' and they go on to say that it has become a 'severe public health concern', with 2013 being a particularly bad year. This raises several questions: is the species more aggressive in its native habitat? Will the invasive Asian hornets in Europe become more aggressive over time (perhaps due to nest density)? Or is the aggression seen in China during recent years simply the result of closer human-hornet contact through massive development projects, which have turned some areas of countryside into towns and cities?

As mentioned before, Asian hornets do extremely well in urban environments (Monceau & Thiéry 2017, Choi et al. 2012), and humans are more likely to accidentally get too close to a nest in this situation. It may even be the case that *Vespa velutina* is forced to nest much closer to the ground in towns and cities due to a lack of tall trees to nest in. Of secondary nests in south-western France, 78% were in natural structures (mainly trees), Franklin et al. (2017). The Association Action Anti Frelon Asiatique (the AAAFA), a French grassroots organisation, lists the dangerous places at ground level or only a metre high that nests have been found in, including under manhole covers, inside gas and electricity meter boxes, in hedges, arbours and fruit trees, and various building cavities. They suggest that an attack can be triggered by strong vibrations and noise (such as use of some garden machinery), and recommend always checking suspect places (even if they aren't actually showing hornet activity) by tapping with a long-handled broom, while being ready to run!

A word about Asian hornets squirting venom

Rumours circulate about Asian hornets being able to squirt venom, but so far I have been unable to verify this, although I have talked to pest controllers and beekeepers in France and Jersey. Traditional hornet suits come with flexible plastic visors instead of mesh veils, but no one I spoke to has seen venom on this faceplate, or seen this behaviour in *Vespa velutina*. However, we should consider that this is a possibility, even if rare. No-one wants to have venom squirted into their eyes. Pest-controllers, who would be most at risk, should take this into account.

Part II
Context

The spread of the Asian hornet

The Asian hornet arrives in Europe

The first confirmed recording of an Asian hornet in France was made in Nérac, Lot-et-Garonne (inland in south-western France) in 2005, by Jean-Pierre Bouguet, an amateur entomologist (Haxaire et al. 2006). When another three Asian hornets were found the following spring, it was suspected that these were probably foundresses too, and it was possible that one or more colonies could have been established in the area for at least a year, putting the invasion at 2004, possibly even earlier. Meanwhile, a nearby bonsai producer had noticed brown hornets flying in the summer of 2004, which he recognised from a trip to China not long before. It is quite likely that this is the origin of the French invasion: a fertilised female could have easily survived the month-long journey by boat from Yunnan, China, hibernating during winter in a box of pottery (Villement et al. 2006). It may be that a single, multi-mated queen is responsible for this spread through Europe, which is unfolding before us (Arca et al. 2015).

Vespa velutina naturally occurs in Asia from Northern India to Eastern China, Indo-China and Indonesia (Carpenter & Kojima 1997), and genetic marker comparisons do support the idea that the French population came from eastern China (Jiangsu/Zhejiang) (Arca et al. 2015).

The speed of spread through Europe has astonished everyone; indeed, the expansion through France has proceeded at around 78 km per year (the pioneer population possibly achieved 30 km in the first year) (Robinet et al. 2016) with some long-distance jumps from the colonisation front of around 250 km, which may or may not be due to human transportation (Rome et al. 2009). This is four to eight times faster than in South Korea, where a similar invasion of *Vespa velutina* began around the same time (2003) (estimated spread 10-20 km per year; Choi et al. 2012). In South Korea, their spread may have been slowed by competition with the six species of native hornets already there (Villemant et al. 2011a).

Map of Europe, showing colonisation, in yellow, by Asian hornets. The initial introduction into France is marked with a red spot. Although there have been several incursions into the UK which have been successfully eradicated, and there is no evidence of colonisation so far, we do not know whether Asian hornets might already be established — hence the big question mark.
This map is based on a map of Asian hornet spread in Europe maintained by Quentin Rome at the Museum National d'Histoire Naturelle (MNHN: National Museum of Natural History in France). It is up-to-date at time of publishing. For a current map, please go to their website:

http://frelonasiatique.mnhn.fr/home

Map from freevector.com

Now (2019), virtually all Departments in France have been invaded, along with northern Spain, two thirds of Portugal, Mallorca and the Channel Islands, with inroads into Belgium, Italy, Germany, Switzerland, the Netherlands and the UK (Rome & Villement 2018).

As far as climate goes, the whole of western Europe seems conducive to invasion by Asian hornets, including Ireland and as far north as Denmark. One model (Villemant et al. 2011a) predicts climatic suitability right up into Scotland and coastal Norway, based on the range of its native territory, but when combined with data from invaded areas, the predicted range is a little more limited.

As Martin (2017) points out, however, although *Vespa velutina* has a massive natural range, in fact this range is occupied by a complex of 12 subspecies: these different subspecies are probably acclimatised to different conditions within this huge range.

Climate change is, of course, another factor to be considered in predictions of spread. Barbet-Massin et al. (2013) looked at climatic suitability for Asian hornets in 100 years' time, based on changing patterns of temperatures, precipitation and seasonality, and found that their predicted range could extend east into central and eastern Europe — to countries which at present have the highest densities of honey bee colonies.

The European hornet at present is found roughly as far north as Yorkshire in the UK (BWARS).

Asian hornets reach the Channel Islands

Sarah Bunker and Bob Hogge

The first of the Channel Islands to detect Asian hornets was Alderney, where they were seen in July 2016. A nest was discovered in the top of a sycamore tree and destroyed. Next was Jersey, with the first sighting made by an amateur entomologist in August 2016. In Guernsey, the first Asian hornet was found in March 2017.

Alderney and Jersey, the islands closest to France, are only 10 and 12 miles, respectively, from the Cherbourg Peninsula of Normandy.

Bob Hogge and a small dedicated team (Monique Pierce, Richard Perchard and Megan Dimitrov) developed their own tracking techniques in Jersey from scratch, relying initially on strong natural history observations. By the time I went out to Jersey in August 2018, a slightly modified version of the original technique was

An Asian hornet feeds on tree exudate (sap or sap modified by fungi and bacteria) among the roots of an old oak, the 'Bee Tree' on Jersey.
Photo by Bob Hogge

finding about a nest a day in a mixture of urban, semi-urban, agricultural and heavily wooded incised valley areas.

In 2017 there were a number of sightings in Jersey, mainly along the east/north-east coastal strip. The first primary nest was discovered in an active beehive in the centre of the Island in April, and this was followed by four more primary nests that were removed and destroyed. The Department of the Environment initiated an extensive trapping programme, using Véto-Pharma traps and bait; the focus of the programme was concentrated on the NE coastal area. Using DNA tests in co-operation with the National Bee Unit in York, the trapped individuals were successful in establishing that there was only one *Vespa velutina* nest in the area.

In July 2017, Bob saw his first *Vespa velutina* feeding from a small depression in a root of the Bee Tree (an ancient coppiced oak with a wild honey bees' nest less than 1 m above ground), which was also attracting other insects, including red admiral butterflies, common wasps, European hornets and various fly species.

A damaged tree like this one can provide a natural feeding station for Asian hornets: they arrive to take advantage of sap oozing from the wound. Photo by Bob Hogge

Having no knowledge on the biology of Asian hornets, the team set about finding out how closely they could work with them, and observed their foraging behaviour.

They started to mark individuals, and found that a queen-catcher worked well for this: once marked they could get times for individuals leaving and returning to the tree. Next they realised that if they could move bait towards the nest, perhaps they could narrow down the area to search, so they set up an experiment to test various foods, and Suterra wasp attractant was by far and away the favourite, and became their bait of choice. Later they discovered a natural attractant even tastier to the hornets than Suterra: camellia flowers.

Once they had moveable bait stations, they began to track in earnest, working out how to get the best views of flight paths and how to negotiate obstacles. Luckily, they found the public, helped by an energetic and sympathetic local newspaper, were happy to let them track the hornets across their land.

By doggedly following the hornets from the first sighting, they finally got a return time for one individual of under 1.5 minutes. The Department of the Environment was contacted and help requested and, while they waited for the cavalry, they continued for another 10 m towards the nest, but then their way was blocked by a tall oak tree, over which the hornets flew. So, with the cavalry in place, they then tried a different strategy, positioning spotters between the blocking trees and shouting "on way" as the hornet passed over. The search for the nest was initiated and, 45 minutes later, two hot and sweaty spotters emerged from the trees with broad grins and it was obvious that they had found the first secondary Asian hornet nest in Jersey, 39 days, and a steep learning curve, after Bob had seen the first worker at the Bee Tree 1.6 km away.

Due to the excitement generated in the press from finding the first secondary nest, public awareness soared, and a couple of urban secondaries were quickly found just by people looking up; one in an oak tree and one in a cavity wall. By careful and consistent use of the tracking method, nine secondary nests were discovered in 2017.

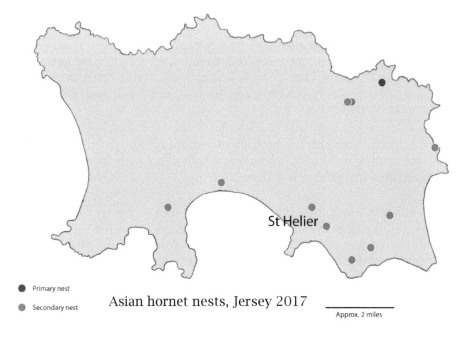

Asian hornet nests, Jersey 2017

Asian hornet context

What the Jersey hornet trackers learned from the first season:

- It is possible to track Asian hornets and find their nests with a minimum of equipment.

- Even without knowing the time they spend in the nest, a good enough estimate of distance from timing the return flight can lead to finding the nest.

- Nests aren't always found in tall trees.

- Placing bait stations in open spaces gives a much better chance at reading the flight trajectory.

- Even later in the year, when Asian hornets are feeding a lot of larvae, they came to a sweet bait (Suterra) much more reliably than a protein bait (shrimps, fish, lobster).

- The vast majority of landowners were happy for hornets to be tracked across their land by a careful and courteous team.

In 2017, four primaries and nine secondary nests were found: in 2018, 14 primaries and 36 secondaries were found.

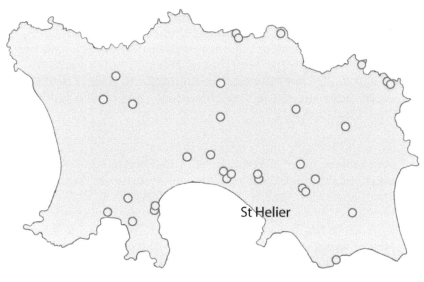

○ Secondary nest   Asian hornet nests, Jersey 2018   Approx. 2 miles

Asian hornets in the UK

For an up-to-date map of sightings of Asian hornets in the UK, with brief details, go to the British Beekeepers' Association (BBKA) website and look for the Asian Hornet Incursion Map (see 'Resources').

The first reported sighting of an Asian hornet in the UK was at an apiary near Tetbury, Gloucestershire, by a beekeeper, in mid September 2016. The nest was located and destroyed a couple of weeks later by the National Bee Unit (NBU). It was found near the top of a 55-foot cypress tree, and the vast majority of the workers were observed foraging within 700 m of the nest (92 of 94 observations within 700 m; another two observations 1.15 km away, feeding on ivy), which fits with other observations on foraging ranges (e.g. Poidatz et al. 2018). Once analysed, the nest was found to contain 70 adults: 57 females (workers) and 13 males. Interestingly, the males were diploid, therefore likely to be sterile. There were also eggs (including those that would have turned into sexual males), larvae and pupae. Genetic analysis showed that the queen had only mated once, which suggested low nest density, and the genetic markers were consistent with these insects being closely related to the French Asian hornets: part of the same European invasion (Budge et al. 2017). To check that this nest was the only source of Asian hornets, more than 220 field inspections of apiaries and flowering forage sites were conducted within a radius of 17.5 km from the nest. No further hornets were found.

Then, the day after the Tetbury nest was discovered, a beekeeper in north Somerset (54 km from the Tetbury nest) reported having trapped a single foraging Asian hornet between April and June (2016), which was now dried up but positively identified. Another massive search of 186 apiaries, forage sites and water sources ensued, within 20 km of the original sighting. No Asian hornets were spotted, perhaps suggesting that the forager was from a primary nest that failed (Budge et al. 2017).

Data from the Tetbury and Somerset hornets suggested that either they were two separate introductions of live hornets into the UK from a genetically related area of low nest density, or that they both came from an earlier, undetected colony (Budge et al. 2017).

Two dead Asian hornets also turned up near Tetbury. One was found in a stack of wood recently imported from the Loire Valley, France, found a week after the Tetbury nest had been removed. The second was in some camping equipment near Bath a week after the wood stack discovery, that had also returned from the Loire Valley.

Asian hornet context

In March 2017, an individual Asian hornet was found in a warehouse in Scotland, and in September a nest was found in Woolacombe after a beekeeper reported a sighting in his apiary. The nest was found in 3 days by NBU staff who used triangulation of lines of sight to locate it. After the nest was removed, surveillance was carried out within a 10-km radius of the nest site, and no further Asian hornets were found.

In 2018 there were nine sightings and four nests found. The sightings began in April, with an Asian hornet found in a cauliflower in Bury, Lancashire. The cauliflower was traced back to Boston in Lincolnshire, but no further hornets were discovered despite 190 km of hedgerows being walked by observers. Another individual was found in August on the Poole-Cherbourg ferry, and then on 31st August a forager was trapped in a home-made monitoring trap in the Fowey area, Cornwall. This led to a full-scale nest hunt and began what was to be an extremely busy month for the NBU and seasonal Bee Inspectors, who were brought in to help. The Fowey nest, found in brambles, was destroyed on 6th September, but more hornets found after the nest was removed led to another nest being discovered only 10 m away in a tree. The two nests, 20 cm and 25 cm in diameter, respectively, contained genetically identical Asian hornets, which could mean that these two nests were a primary and secondary of one colony.

While this was going on, a single male was caught in an apple juice trap in Liskeard, Cornwall: when the genetics were looked at later, it seems probable that it was from one of the Fowey nests, although the distance between the two sites was at least 12 miles. No further hornets were found at Liskeard, but a dead Asian hornet was discovered in a house in Hull, East Yorkshire, possibly having travelled from France in camping gear.

On 22nd September a member of the public found an Asian hornet on fallen garden apples in New Alresford, Hampshire, and reported it to the NBU. A nest was quickly found low down in a bush. It was 25 cm in diameter and contained all life stages, including 122 adults. Another sighting a few days later, again on fallen garden apples and again in Hampshire, at Brockenhurst, resulted in another full-scale search for the nest, which was difficult to find in heavy woodland. Tree climbers were brought in to try and get a glimpse of the nest from a different vantage point, and a UAV (unmanned aerial vehicle, commonly called a 'drone', but this causes confusion when beekeepers are involved, as male honey bees are called drones) fitted with an infra-red camera was also deployed, but was unable to find it. Dr Peter Kennedy (Exeter University), who has pioneered radio-tracking of Asian hornets, was also called. He was able to attach a radio-tag to a particularly large hornet and, by tracking it, was further able to reduce the search area until the nest was finally spotted by a volunteer. The nest was 20 cm in diameter, had 68 adults and all life-stages except eggs. Interestingly, the nest looked as if it had been recently damaged as it had a chunk missing: this may have been why the UAV was unable to find it — perhaps it was losing too much heat to be able to be seen with the infra-red camera. Both Brockenhurst and New Alresford are fairly close to several ports: Poole, Southampton and Portsmouth (all around 20 miles away).

In October, a dead Asian hornet was discovered in a workshop on an industrial estate in Guildford, Surrey: it might have arrived on a car transporter.

Finally, also in October, two male Asian hornets were found in Dungeness, Kent, one foraging on ivy in a garden and another at the Dungeness Bird Observatory. In this case there had been strong winds from the south, so the hope is that they were blow-ins from France, and not the result of a local colony.

All the nests so far destroyed in the UK were reported as having been caught before the release of sexuals (potential queens and fertile males). From genetic analysis it is clear that these hornets have come from Europe, rather than being completely new introductions straight

The Brockenhurst nest. The Asian hornet fitted with a radio-tag approaches the nest: inset shows the tag (yellow) hanging under but close to the hornet; the fine wire aerial trails behind. Note that the nest appears to be damaged: normally all the comb is protected by the paper envelope of the nest, and the small entrance hole is half-way up the side in a secondary nest like this. Photo by Peter Kennedy

from Asia. Genetic comparisons show that the nests found in 2018 were not from Tetbury or Woolacombe, which is good as it means that Asian hornets are less likely to have established in England so far; but it does show how many incursions are happening from the continent, and this rate may increase as the density of Asian hornet nests throughout France reaches carrying capacity over the next few years.

It is interesting that the nests discovered in 2018 in England were all fairly small (20-25 cm), about half the size of an average Jersey nest. Whether this is due to them establishing late, or perhaps struggling to expand rapidly, is unknown.

So, from 3 years of incursions we have learned several things: (1) Asian hornet sightings can occur anywhere, due to their arrival by accidental transportation; (2) Asian hornets can successfully build colonies in the UK when a mated queen is present; (3) nests can be found and destroyed, and (4) it is down to the vigilance of members of the public and beekeepers to spot Asian hornets.

The NBU will carry on investigating sightings and removing nests for the moment. As incursions are likely to increase (due to higher densities and wider spread of Asian hornets in mainland Europe), more incidents will probably have to be dealt with year-on-year. Once the NBU becomes overstretched in dealing with cases, there are still the Asian Hornet Action Teams (AHATs) who are ready to help with tracking and surveillance. In this way, it may be possible to keep them at bay for some time. However, it will only take one nest to remain undiscovered for them to get properly established; and a year or two after that happens it will be nigh on impossible to eradicate them.

One paper (Keeling et al. 2017) has attempted to model the spread of Asian hornets through the UK based on establishment at Tetbury. The authors show that, without control, it is possible for *Vespa velutina* to become established across most of England and Wales in 20 years (petering out along a line roughly from Newcastle to Carlisle) — unless our temperate maritime climate is a strong deterrent. In this scenario, although moderate detection and destruction of nests would slow the spread, extremely high levels of detection would be needed to eradicate Asian hornets.

The sobering thought is that Asian hornets may already have established in the UK, but have not been noticed.

Asian hornets on a primary nest: some are involved with nest construction (see darker band of papier-mâché which is still wet), while others patrol the surface. Photo by Francis ITHURBURU (CC BY-SA 3.0)

Asian hornet context

A National Contingency Plan

In some ways, Asian hornets fit into the same kind of planning scaffold as other biological threats to the UK, including diseases of humans and animals. As you would hope, the British governments of England, Wales, Northern Ireland and Scotland have contingency plans in place and, considering the steady invasion of France, have had time to hone these plans and put some into practice since the first sighting of an Asian hornet in Tetbury, Gloucestershire, in 2016.

The Asian hornet is classed within the European Union as an Invasive Alien Species (IAS) of Union Concern, because of its rapid spread through member states and across borders. An IAS is an alien species whose introduction or spread has been found to threaten or have an adverse impact upon biodiversity and related ecosystem services. According to the International Union for the Conservation of Nature (IUCN), IASs are the most significant threat to biodiversity after habitat loss. To try to counter this invasion, member states are obliged to try to eradicate it, and when that fails, to manage the pest.

The full NBU (National Bee Unit) contingency plan will download automatically from:

www.nationalbeeunit.com/downloadDocument.cfm?id=675

Here, I would like to summarise how incursions are dealt with.

The main players

The Pest Specific Contingency Plan for the Asian hornet has two pages of acronyms in the glossary! I have sketched out the main players (diagram) to give an idea of how they are related. Government organisations have their own top-down hierarchy, intended to deal with pest incursions to protect the public, livelihoods and the environment. Into this existing setup, the bottom-up organisation of AHATs (Asian Hornet Action Teams) has sprung, at present as part of local beekeeping associations, with a single point of contact via the British Beekeepers' Association (BBKA) with the National Bee Unit (NBU).

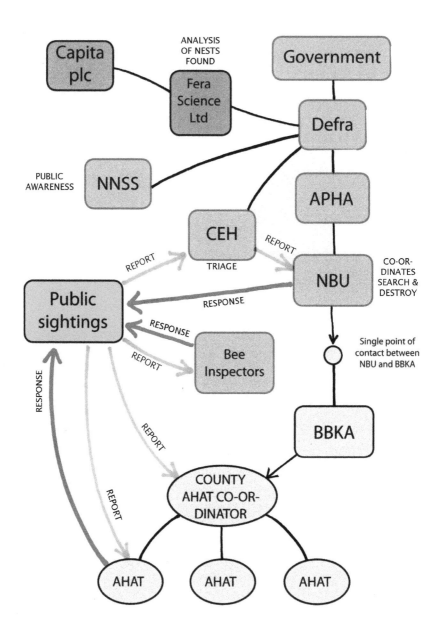

Diagram of main players involved in Asian hornet sightings and responses. Acronyms: Defra, Department for Environment, Food & Rural Affairs; APHA, Animal and Plant Health Agency; NBU, National Bee Unit; NNSS, Non-native Species Secretariat; CEH, Centre for Ecology and Hydrology; BBKA, British Beekeepers' Association; AHAT, Asian Hornet Action Team

Waiting…and spotting

In 2011, the Non-native Species Secretariat (NNSS) were charged with putting together a risk assessment for *Vespa velutina* (Marris et al. 2011). In it, they tried to find out as much as they could about the insect and work out how it could get into the UK, and which paths were most and least likely. They concluded that the arrival of *Vespa velutina* into the UK was very likely, as was its rapid establishment and spread. Less certain at the time was its impact, due to lack of reliable data from mainland Europe.

The first risk to manage is the arrival of *Vespa velutina* in the country, but with the amount of traffic between the UK and mainland Europe, and the possibility of Asian hornets also making it across the Channel under their own steam (or helped by strong winds), this remains something that cannot be dealt with. And, judging by the scattered sightings so far, the risk assessment report was correct in reckoning that *Vespa velutina* could turn up anywhere, via any one of several pathways.

Already in place is a network of sentinel hives all over the country, and especially near ports and airports, which are managed by beekeepers on behalf of the NBU. They are constantly looking for possible bee pests, such as small hive beetle (*Aethina tumida*), which is expected to turn up at some point.

In addition to the beekeepers with the sentinel hives, everyone else in the country can act as the eyes of the NBU.

At the moment we are in the eradication stage, and if incursions are spotted early enough each year, there is a real possibility of extending this phase for some time. To be able to do this, a massive pubic awareness programme needs to be rolled out. So far, there has been little national media coverage, and much misleading information in the press and online.

The NNSS has put out two posters: one for identification, and one a 'Wanted' poster — the NBU is happy to supply you with these.

Beekeepers are probably the most informed about Asian hornets at the moment, because of the threat to honey bees. There have been quite a few articles in beekeeping magazines, and Stephen Martin's book *The Asian Hornet: Threats, Biology & Expansion*. But it is important to emphasise that this isn't just a beekeeping problem: *Vespa velutina* is a voracious predator of other pollinators as well as many other small creatures, including flies, caterpillars and spiders.

Biodiversity in the UK is in such a sorry state that this new invader could do real damage to our ecosystems.

Getting the wider public up to speed might be largely down to AHATs (see next section), with the NNSS targeting all sorts of organisations connected with the natural world, community groups and outdoor leisure.

Meanwhile, the NBU has been planning and practising its response to an incursion, and over the last 3 years the preparation has turned into the real action of finding and destroying nests.

Any sighting, with a photo (if possible: a good written description will be taken seriously), location and date of sighting, and contact details should be sent direct to:

alertnonnative@ceh.ac.uk

If you are able to catch the hornet, so much the better: if you can get it to feed at a bait station, you should be able to catch it in a jar. Although they can and do sting, they are very docile when away from the nest and being fed.

Suppose you know your insects and were able to get a photo of something you are 99% sure is an Asian hornet (you've never seen a live one before). You send in your report via email or the app, and it ends up in the in-box of the CEH — what happens next?

The CEH received 8000 possible sightings last year (2018). Out of these, half could be identified (obviously the vast majority of these weren't Asian hornets — a reply is sent back to the person who sent the report in, telling them what the insect is): a positive identification or a credible sighting of an Asian hornet gets sent immediately to the NBU, triggering their response.

The problem is the other 4000 unidentified insects. For these, an email goes back to each reporter with links to further identification sites, and encouragement to get back with a photo/a better photo, or a specimen. How many of these are indeed Asian hornet sightings? This is potentially a big problem, and one that, again, local AHATs may be able to help with by talking through the sighting and by setting up a monitoring station.

Dealing with an isolated incident

When the NBU receives a positive or credible sighting of an Asian hornet (which always seems to happen on a Friday afternoon!), their plan swings into action, with a local Bee Inspector dispatched to see whether live Asian hornets can be located. Then the whole team moves to the area, initially setting up widespread surveillance and then moving on to track insects back to their nest using a method similar to that later developed in Jersey (there is also the possibility of using radio telemetry: see 'Control: Radio telemetry').

Once a nest has been located, the colony is destroyed at dusk, and the nest is removed the following morning for analysis.

Surveillance then continues, to make sure that there was only one nest involved in the incident (this is how they found the second nest at Fowey: the second nest was only 10 m from the first). If there is any possibility of sexuals having left the nest, surveillance with monitoring traps would be continued intensively in an area with a 20-km radius from the nest, and neighbouring apiaries monitored for at least a year.

Some sightings are of individuals that have hitched rides and are not part of a live colony: early in the year they may be hibernating queens that have ended up in the UK. Later, lone individuals may be found, which have been accidentally transported to the UK. These sorts of reports are dead-ends; unable to find any more Asian hornets, there's not a lot more you can do except to keep checking monitoring traps. If there are more Asian hornets around, they will find the traps.

This approach for detecting and tracing Asian hornets, used in 2017 and 2018, will be carried on into 2019. The NBU has said that they will not change tactics during the Asian hornet season, so if there is a change in the campaign, it is likely to be in late autumn, after sightings have stopped for that year.

The NBU will try to keep on top of these isolated incidents for as long as is feasible, using AHATs for help with surveillance. Seasonal Bee Inspectors are now employed for an extra month each year (October) to help out. However, more than one nest at the same time will start to stretch the NBU, and a few being discovered in different parts of the country will be very difficult: it may be at that point that AHATs could get permits to catch and release hornets and so help track them.

Losing control

If a nest goes undetected, and the sexuals emerge, mate, and the new queens disperse, then population growth can be explosive: according to the NBU, one incursion could lead to 20 or 30 nests the following year, but this is probably a worst-case scenario. In 2018, the nests found in England were much smaller than those regularly found on the continent or Jersey. Because there is a direct correlation between nest size and population, smaller nests should produce fewer males and gynes, leading to a smaller number of nests the following year than you would get from a large nest. However, the small sizes of the nests could be due to a very late spring or late-starting queens.

We still know nothing about dispersal of mated gynes (queens) — we just know that the spread in France proceeded at a rate of 60-100 km per year; probably a mixture of gyne flying dispersal and accidental transportation.

At some point, unless there is a game-changer (like an extremely effective pheromone trap which drastically reduces the numbers of Asian hornets in France and therefore the likelihood of incursions into the UK), there will be too many nests to deal with individually, and the NBU will switch strategies from eradication to management.

The 'new normal'

In a management scenario, it is not yet clear who will pay for destruction of nests, but presumably local authorities would deal with nests that are deemed to be a particular public nuisance — for example around schools and hospitals. We would all have to learn to live with a potentially very dangerous species (accidental disturbance of nests being the most dangerous thing), and if urban areas are colonised, we would be coming into contact with *Vespa velutina* more and more. Beekeepers would have to defend their hives and perhaps deal with nearby nests.

If Asian hornets become established, it will be very difficult to get rid of them. The longer we can hold them off, the more possibilities might arise from pheromone-based attractants and targeted poisons, possibly allowing their numbers to be kept down long term.

The birth of Asian hornet action teams

Colin Lodge

In September 2017 a beekeeper named Martyn Hocking made a startling discovery in his apiary in Woolacombe, North Devon. Flying around his hives were dark coloured insects with an orange band on their abdomens. He knew what they were but could not believe it; he reported them to the authorities only to be told Asian hornets are not native to this country so they were unlikely to be Asian hornets. He finally gathered enough evidence to convince the National Bee Unit (NBU), and Woolacombe was invaded by bee inspectors and other Defra personnel.

Unfortunately, Martyn was left out of the loop during the NBU's search-and-destroy campaign, even though he had discovered them and they had been attacking his bees. It was this 'Men in Black' approach to Asian hornets by the authorities at the time that caused a bit of a rift between beekeepers and the organisation that is supposed to support them, the NBU.

Jill and Ken Beagley suggested and organised a meeting of Devon beekeepers to discuss what could be done about this threat, and this meeting was held on 20th January, 2018, in Harberton, North Devon. At the meeting, Nigel Semmence, the NBU's Contingency Planning and Science Officer, was asked whether beekeepers could be co-opted to work alongside Bee Inspectors — to which suggestion he gave a flat refusal. He told the audience that only NBU personnel would be allowed to work on tracking and extermination because only they had the authority to enter property and only they knew how to track and destroy Asian hornets.

Martyn gave an impassioned plea for beekeepers to be involved in the fight against Asian hornets: as his last statement of his talk he said:

> *If the Asian hornet becomes listed as being established, without major changes in the policies of Defra, FERA, CEH, APHA and NBU, then history may yet decide that they were part of the problem rather than the solution.* *

---

* See Glossary or diagram on page 60 for meanings of these acronyms.

At the end of the Harberton meeting I was able to outline my ideas for Asian Hornet Action Teams. Torbay Beekeepers' AHAT had been formed in the November or December and we had been developing an approach and aims since that time. Initially, AHATs were conceived to offer more manpower to help with incursions, to reassure local beekeepers that something was being done, and to teach people to identify Asian hornets.

Newton Abbot Beekeepers' Association (BKA) and Totnes and Kingsbridge BKA soon set up AHATs along the lines of the Torbay model and we came together to form the Home AHATs. Members of these AHATs formed a steering group for developing the AHAT movement in Devon and to oversee its adoption by Devon BKA (DBKA).

The AHAT website was the suggestion of James Schindler-Ord, and is managed by Mike Ticehurst. It has a reporting facility, and directory of AHATs. The BBKA is now maintaining a map-based list of AHATs on its website, and all our AHATs will be added in time. At present we have about 35 listed, but advice from BBKA to set up a team in every branch or association in the land should increase this number enormously.

Throughout the summer of 2018 we provided the assistance and reassurance the public needed, especially by identifying suspected insects (all of which turned out to be *Vespa crabro*).

The NBU is acutely aware that it could and will miss some occurrences of Asian hornets because it does not have a large enough workforce to carry out the necessary investigative work. We can and must provide that support.

A vital task for AHATs is in educating the public in Asian hornet identification. We need Asian hornets to be discussed on TV, on the radio, in newspapers and other publications. We need a similar spread of identification posters as happened in the past with the Colorado beetle, a serious threat to potatoes, which older readers will remember.

In the end it boils down to a question of deciding whether we should try to do something. It has always been my contention that if we do nothing we will bitterly regret the lack of effort. As we have decided to try, the effort must be whole hearted and any dissent from this position will seriously undermine not only our efforts but the effort of the Inspectorate who, whether we acclaim their effort or not, have decided that prevention will be better than curing, since curing this problem will be nigh on impossible.

Asian hornet context

> ## The fourfold aims of Asian Hornet Action Teams
>
> 1. To increase the numbers of people working on each case of incursion to assist Bee Inspectors, who are thin on the ground
>
> 2. To be ahead of the game in that AHATs will initiate positive identification of insects in instances of Asian hornets being reported in a locality, and commence efforts to locate nests if this is needed by the NBU
>
> 3. To give local beekeepers and the public some reassurance and confidence that Asian hornets are being dealt with in the speediest way possible
>
> 4. To educate BKA members and the public in the identification and risk of *Vespa velutina*, as our colleagues in Jersey have done successfully in 2018

You can find your nearest AHAT by going to the AHAT website, or to the AHAT page on the BBKA website (see 'Resources').

www.ahat.org.uk

Part III
Control

Finding Asian hornet nests

Introduction

It is important to reiterate that at the time of writing (spring 2019), there is no evidence that Asian hornets have established in the UK. The correct procedure if you think you have spotted an Asian hornet is to try to get a photo, or the actual insect, and report it (see back cover). If it is an Asian hornet, the NBU will look for and find the nest, if there is one, then destroy and remove it.

Also at the time of writing, the Animal and Plant Health Agency (APHA) does not allow the release of an 'alien invasive species' (and *Vespa velutina* is classed as that) without a permit. This is understandable in that this proscription is a one-size-fits-all law that would apply equally to any invasive species (plant, animal etc.) and exists to prevent the spread of the non-native species. However, in a scenario where a colony of Asian hornets has established and a steady supply of workers are visiting a bait station, releasing a marked hornet in order to find the nest and destroy it would seem to warrant an exception to this rule. Not only are you helping destroy the nuisance species, but you are releasing a worker that cannot found its own colony, and doing it within its existing home range. Also, in a few minutes it will even come back to you!
Consider another situation, however. You find a single, live Asian hornet in the spring which you mark and release. You didn't realise it, but it was a foundress (queen), and she doesn't come back: perhaps she was feeding during her migration. There was no nest to find, and now you have lost what would have been valuable evidence of an Asian hornet in that area, and a valuable resource for the NBU to look at the genetics of. She then goes on to found a colony which is undiscovered and leads to the establishment of Asian hornets in the UK. *If only you had caught it and reported it.*

Asian hornets could establish in the UK at any time, and I want you to be prepared. However, it hasn't happened at the time of writing, so please follow the national guidelines and don't track Asian hornets unless the national policy changes. I have written the following section on tracking as if a permit has been granted in order to make it easier to read and absorb.

Note: Although the majority of **primary** nests (the small nests founded by the queen in the spring and often found in man-made structures) are simply found by curious and informed people spotting the nests themselves, there is no reason why early workers cannot be followed back to their nest. Indeed, nests might be easier to locate earlier in the summer because there is evidence that the workers don't fly as far from the nest (foraging range around 300 m, Sauvard et al. 2018).

The Jersey method

Developed in Jersey by Bob Hogge and a small, enthusiastic team, this technique has been refined throughout 2018 by the many volunteers tracking around the island. In Jersey, permission to release insects was obtained from the Jersey States Department of the Environment (Jersey is not part of the UK and not fully in the EU). Because it relies on people power and a few simple pieces of equipment, it can be practised by anyone who is methodical and has a calm disposition towards insects, especially the stingy sort. And I'm not talking about hornets here, but common wasps, because you may be working in a cloud of them (they love Suterra wasp attractant).

Etiquette

At all points, tell people who you are, what Asian hornets are, and explain what you want to do. If you are from an Asian Hornet Action Team (AHAT) you may have an information leaflet to give to the land owner; at the very least you should carry some Asian hornet identification leaflets. Get permission from the landowner to go onto their land. This is essential. All you have to do is knock on the front door of a property or call out in a farmyard. If you can't raise anyone, go back later, or try a neighbour who might know where the landowner is. If you don't get permission, either through not finding the landowner or them refusing permission, then try and get permission for neighbouring land instead — you should still get times and flight directions, and in urban settings you can easily carry on tracking.

- Always leave a contact number so that the landowner can get back to you.

- Never leave unattended bait stations where children or animals could get stung.

- Be considerate when parking.

- Label bottles of wasp attractant, and don't leave them lying around — it looks like fizzy pop.

- Make sure that bait stations are removed when you have finished.

- And of course follow the country code if that's where you are hunting.

Getting set up

Starting from the point of getting a confirmed sighting, the first thing to do is set up a bait station where the hornet was seen (see Box 1). If you know you can return later (even the next morning) then it is ideal to set up three or four bait stations at and around (approximately 100 m from) the original site. Then, by the time you get back, some of the bait stations should have visiting hornets, saving you a lot of time.

As soon as you know where you will be searching, print a Google Map, if you can (or an OS aerial map, if you are a subscriber). Aerial or satellite maps are incredibly useful because they show actual trees, hedges, buildings and roads. If the position of your bait station is in the centre of the map, you will want about a kilometer of territory showing on the horizontal axis (500 m on each side of bait station, east and west). Because most computer displays are landscape (rectangle with its long side at the bottom), the amount of territory in the north-south directions (up and down on the computer screen) will normally be quite a bit less. If you can print your map on an A3-printer, you can experiment to include more territory without the landmarks getting smaller. This is highly recommended.

What you will need

Essential for noting flight direction: map, bait station (including extra attractant)

Essential for timing: queen catcher, uni POSCA marker pens, some kind of timer

Very useful: binoculars, compass (if you know how to use one), stopwatch (they are simple, robust and can be operated easily with one hand), walkie-talkies, notebook and pen for noting times and plotting on map

Good to have: food, drink, light camping stool (you will need to wait patiently for returning hornets and this can be a comfortable option)

BOX 1. Setting up a bait station

Just so that you can be clear with your fellow trackers, we are adopting the following terminology: the bait station is the one that you set up using an attractant in a dish. A feeding station is a place where the hornets are naturally feeding, whether for protein (e.g. at a hive or carcass), or for sweet carbohydrates, like camellia flowers or ivy.

The best place to site a bait station is somewhere fairly open where you will have a chance of seeing the direction in which the hornet flies off. *Vespa velutina* vanishes against a complex or dark background, especially hedges and trees, but also houses and rooftops. The more sky you can see, the better. Elevated sites are particularly good because the route to the nest will have fewer obstacles and the flight path is likely to be straighter. Although they will often fly along landmarks such as hedges, walls and even roads, they do not have trouble picking out a bait station in the middle of a field as long as they pick up the scent of the attractant. Obviously keep the bait station away from the public and animals.

The bait station itself should also be slightly elevated — it makes it easier for both humans and hornets to find. It also makes it much easier to observe and catch insects. Cheap yellow builders' buckets from a DIY store are great — you can see them from a long way off and they can be used to carry things. Turn one upside-down and place a shallow container on top (flower pot saucers are good, but need weighing down with a stone if it's windy). Next some tissue, kitchen roll etc. and then pour some attractant on top. The aim is to soak the paper and to have no more than 1 mm of liquid in the bottom of the container. Some people put a stone on the tissue as a matter of course, to keep the tray in place and to give a non-sticky landing area for the hornets; a disadvantage of this is that sometimes the stone gets in the way of catching a worker. Another tip is to put the bait station in the sun when possible to warm and diffuse the scent.

Suterra is a wasp attractant that works extremely well for Asian hornets. But it does attract wasps, sometimes dozens of them, so you need to be OK with a lot of wasp activity around you. Both the wasps and the hornets are very focused on feeding, but you need to watch out for the wasps when you get your sandwich out! The perfumes in sunscreen and other products can also be attractive to wasps, which can be distracting. The wasp attractant that comes with Véto-Pharma traps has been seen to attract honey bees, so should be avoided.

You can also try the home-made attractants that people use for Asian hornet traps in France. Here is an example: equal quantities of biere brune (a sweet malty beer), white wine and red fruit syrup [this refers to the syrups available in France which are diluted to make soft drinks, like our 'squash', only much thicker and sweeter; they suggest syrups such as grenadine (pomegranate), blackcurrant or strawberry. You could try Ribena, jams, or fruit syrups used on ice cream or in milk shakes. The alcohol is supposed to repel bees.

You can also use a protein bait. Raw fish fillets and shrimps have been used successfully. The NBU observed that, when given the choice of sweet bait and protein bait, individual Asian hornets would stick with one or the other, over a days' tracking. In Jersey, Suttera is used because it doesn't go off quickly and is therefore less work (cleaning, replacement). Different baits may be effective in different locations — it depends on what they have found rewarding previously.

It is good practice to leave a laminated information sheet next to a bait station, even if you don't expect anyone to come across it.

A well set-up bait station showing the amount of Suterra to use. There is one marked Asian hornet: dead wasps have probably been killed by Asian hornets. Photo by Judy Collins

Catch your worker

A team of six people is ideal, then you can work with two people at each of three bait stations and communicate by walkie-talkie.

Immediately you arrive, you can look for the flight direction of hornets that are taking off from the bait. Ideally, use a compass; otherwise note carefully the landmark that best shows where you saw it disappear. A distant landmark will give you a more reliable direction to record than a closer landmark will. Incoming hornets are very difficult to spot before they are quite close to the bait.

Stand well back from the bait station unless you are actually catching, or you may obscure the hornet's view and change the landmarks she is looking for. Leave any hornets alone until they have settled and are feeding; let them feed for half a minute before you try to catch one. To get an idea of flight direction, any hornet will do, but to time a hornet you will need to mark it: most people can't tell them apart!

I know — after what you have read about these insects, the thought of catching one without a 6-mm-thick full-body hornet suit sounds scary, but in reality, when they are a long way from the nest and totally absorbed in re-fuelling, they are quite docile. You do not need protective clothing against hornets when tracking, but you might need protection from nettles and brambles.

Marking an Asian hornet using a queen-catcher and uni POSCA marker pen. Photo by Sue Baxter

A piece of beekeeping equipment called a queen catcher is the easiest thing to use to catch your hornet (see photo), the plunger-type is simple to use; you can get them from beekeeping suppliers. Move up to the bait dish gently and slowly and carefully place the tube right over the insect, making sure you don't trap its legs, and wait for it to start walking up the inside. Don't chase the insect around with the tube like I did on my first attempt: you'll just spook it.

Keeping the tube vertical (with the opening at the bottom) to take advantage of their natural desire to move upwards, lift the tube and swiftly pop the plunger into place, moving it gently up the tube until you have the worker pinned against the grid at the top with her top side ready to mark. If she's the wrong way up you will need to back the plunger off and let her move around until she is standing on the foam of the plunger and try again, using just enough pressure that she can't wriggle around. Now you can use the uni POSCA markers through the grid to mark the thorax (between the wing bases) and perhaps the abdomen (bulbous back half). As soon as you have marked her, lower the plunger a little so that she can move and the mark can dry. After 30 seconds or so (to allow the ink to dry), lower the plunger and re-introduce her gently to the bait so that she can have her fill and leave when she is ready. She can get aggressive with the plunger, ripping off good-sized chunks of the foam, but she does not remain aggressive when replaced gently onto the bait station. See Box 2 for information on colour coding.

If you need to mark an individual without catching it first, let it feed for at least 30 seconds and then move in very slowly and gently with liquid Tippex on a foam or brush type applicator. One or two dabs to the thorax or abdomen should give you enough of a mark to distinguish that individual. Avoid getting ink on the hornet's head, eyes, wings etc.

Stand back, make sure your stopwatch is zeroed and you have a good view, ready for take-off.

Off she goes!

Experienced trackers begin to get a feel for when a hornet is about to leave. She is less focused on feeding and may start to clean her antennae or legs. Some appear to give no warning. There she goes! Hit the stopwatch button!

Getting a visual fix for direction of flight is so much easier with two people. Although it may seem pantomime-like, the two of you pointing at the hornet with outstretched arms, and giving an out-loud commentary on its position can really help. She may make a couple of orientating circles above the bait, then very swiftly she will head off, often becoming quickly lost against a background, or disappearing over a house.

Where there is a clear sight over a good distance, you might be able to watch the flight itself with binoculars. In open fields you can follow a hornet against the sky until it is too far to follow. Using binoculars for this needs a bit of practice: stand a short distance back from the bait, facing the direction of flight. Focus the binoculars on the hornet on the bait and wait for her to fly. Here's the tricky bit: the first few seconds of flight needs very rapid adjustment of your binoculars' focus, so be ready with your finger on the focus wheel and pull it round fast while following the hornet. If you can keep it in sight for those early seconds it is quite easy to follow after that, because the focus changes only gradually once it has reached a certain distance away.

Fix the point of disappearance in your mind. A notch in the canopy of trees? To the left of a distant block of flats? Mark it on your map. Take a bearing if you know how to use a compass. Then wait for her to come back, staying focused on the dish so you don't miss her arrival.

When she gets back, stop the clock and note the time.

We're going to use a rough estimate of every minute spent away equating to the nest being 100 m away. This rule of thumb was developed by John de Carteret who, by carefully plotting bait stations and nest sites on Google Maps, then comparing times given to him by trackers, noticed that this was a pretty reliable figure to use.

So if your first time is 5 minutes 37 seconds = roughly 5.5 minutes = roughly 550 m to the nest.

Don't just go with the first time you record. Try and get consistent results, which may involve observing half a dozen flights. Out of your roughly consistent flight times, the quickest time should be the most accurate (see Box 3).

Keep on tracking

Once you have consistent times and a consistent direction, it's time to move your bait station in the direction of the nest, close to where you saw the hornet disappear — but again, try not to get hemmed in. For example, if she is disappearing through or over a hedge, and the land is open on the other side, then station someone on the other

BOX 2. Colour coding your hornets

When you are working as a group, tracking in one area, it is useful if each tracker/pair of trackers have their **own colour for marking the thorax** of all the hornets they catch. After marking the first, add a colour (any colour: the paler ones show up best against the black of the hornet) to the abdomen of the next, so the hornets can be recognised individually. For each, record the colouring, place, date and time.

When you move your bait station (hopefully towards their nest), you can see whether they continue to visit, or whether you start intercepting hornets from another bait station. Experience suggests that workers choose the nearest bait they can find and abandon more distant baits.

Knowing where you first marked a hornet can help you/the team to understand the range and flight-line of workers from a nest. One blue thorax and pink abdomen worker faithfully returned to a moving bait station over 3 days. The pictured hornets were probably gathering soft bark to extend their nest, which the blue and pink worker led us to find, 610 m away.

Photo and text by Sue Baxter

side and see if you can spot her coming through or over when your partner shouts that she has left the bait station. If she is still on track on the other side of the obstacle, then put the bait station on the other side where you have more of a chance of seeing the flight path.

When the bait station is moved, it can take a while for the hornets to find it again. Sometimes they go back to the original place and then follow the scent to the new location. Sometimes they find it straight away, but it can take a while.

Repeat the process and keep closing in. In a straightforward case you can continue until you get the time down to a minute or less, when you can start to search for the nest.

More often, though, you will be faced by some kind of obstacle — a row of houses, a wood, a field with animals in — and you will need to work around. If you have a solid idea of the distance the nest is at (from consistent "return to bait" measurements), your best manoeuvre is to try and triangulate by placing two bait stations beyond where you think the nest is. The more equilateral the triangle of bait stations, the more accurately you can pinpoint the nest. Make use of any good vantage points. If you can't triangulate, then try to at least put a bait station beyond where you think the nest is (back stop), and make sure the hornets are flying back towards the nest.

Use your map to draw lines from these new bait stations in the direction that the insects are flying — you are now closing in on the nest using three intersecting bearings — theoretically, where the lines cross is where the nest is!

Catch and release from a better place

If you get stuck in a position where it is very difficult to see a flight path, you can use a technique that can avoid having to set up a new bait station and wait for regular visitors. This technique is not quite as dependable as using a new bait station, but if it works you can really cut down the tracking time.

Once a hornet has landed and has been happily feeding for at least 30 seconds, gently cover it with a fairly large container (the container of a Véto-Pharma trap is ideal, or a plastic pint glass with some small holes in it), pick up the dish and container, holding them together so that the hornet can't escape, and walk gently to an open place where you have a much better view. Put the whole lot down (if your team-mate has brought the bucket, so much the better), make sure the hornet is still feeding and then gently remove the container and back

BOX 3. Flight times and directions — confounding factors

Various things can slow hornets down en route to and from the nest:

Complex route — Asian hornets seem to navigate by using landmarks, so may not take a direct route. They have been seen flying along hedges, walls, roads and valleys, and then turning corners to another landmark or the nest. Obviously this behaviour will affect time and direction. With repeated visits to the bait, they may develop a more direct route as they learn the way. The return times will then reduce and give a more accurate estimate of the distance to the nest.

Resting/cleaning — Return times are also much delayed by hornets pausing after leaving the bait station to spend time on a nearby branch before setting out for the nest. This can occur when their feet are sticky and need to be cleaned, particularly when they have walked up and down on the sticky paper trying to find wet bait. A clean, dry stone in the dish allows the hornets to use the sides of the dish or the stone to sip from. Jersey trackers think this is crucial for getting accurate travel times.

Wind — Depending on strength and direction, this can slow or speed up different sections of their journey.

Obstacles — They tend to go around or over obstacles, making detours that will make their journey longer.

Time at nest — At the nest they presumably offload some of the bait liquid they have carried back. It is not known how long this takes, but if we assume it takes a similar time each trip then, as you get closer to the nest, this offloading time becomes more significant and a bigger proportion of the time spent away from the bait station.

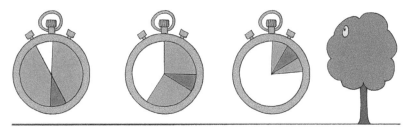

As you get closer to the nest, the hornet spends less time flying to and fro (orange) and the same amount of time unloading at the nest (red)

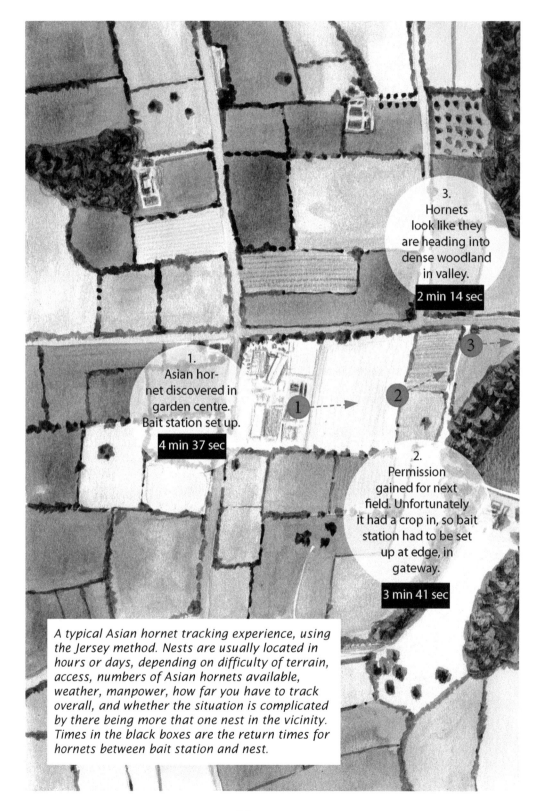

A typical Asian hornet tracking experience, using the Jersey method. Nests are usually located in hours or days, depending on difficulty of terrain, access, numbers of Asian hornets available, weather, manpower, how far you have to track overall, and whether the situation is complicated by there being more that one nest in the vicinity. Times in the black boxes are the return times for hornets between bait station and nest.

Asian hornet control

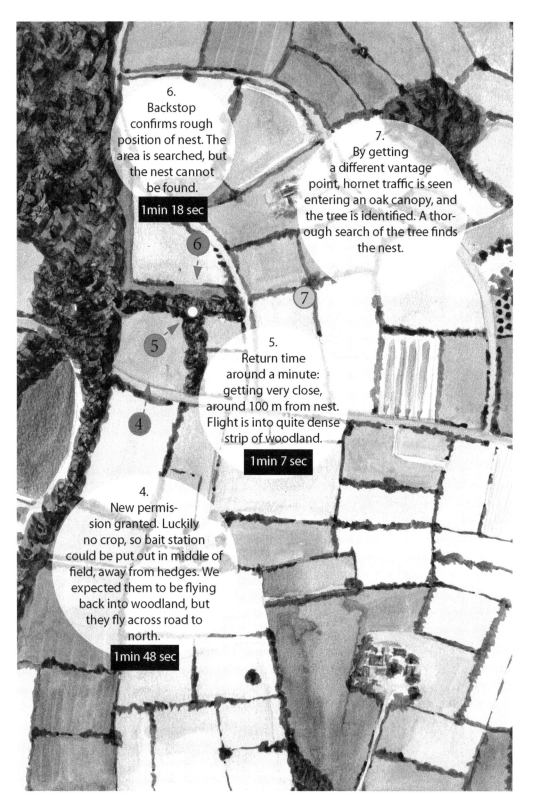

off. Hopefully she will carry on feeding for a bit before she takes off. Because she is in a new position, she will circle a few times when she takes off, trying to get her bearings before flying to the nest. If you are lucky, you will be able to make out the new flight path straight away. This technique is more likely to be reliable if you repeat it several times, with the same or different hornets, as they sometimes fly off in an apparently random direction.

Attaching a feather

The hornet hunters in Asia have long used a technique of attaching a thread to a hornet, with a small feather attached to it, which makes it much move visible as it flies off (Martin 2017). Nigel Errington developed the following technique using fly-fishing materials (Box 5). This technique can be especially useful once you get the time down to under a minute.

Looking for the nest

Many hours can be wasted by starting to look for the nest too soon. You can see a really promising looking clump of trees, or a particularly large tree, and part of you thinks, 'I could short-cut all this slow, patient bait-moving and heroically find the nest, based just on a hunch...'. OK then, try it — I'll see you back here by the bait station in an hour! Honestly, you will probably want to try this, but after a couple of goes and hearing your team buddies on your walkie-talkie patiently discovering that the hornets are flying away round a corner, you will come round to the methodical approach, which wins out every time.

It is best not to start actively looking for the nest until you are getting return times of under a minute, or at least a convincing estimate of nest position from three crossing bearings. Get as many people searching as you can — often a fresh pair of eyes can spot a nest in a tree already searched several times!

Nests are notoriously difficult to spot from directly under a tree. It may be better to be on the look-out for 'traffic' (see later) to guide you to a particular tree, and only then spend time looking up at tree tops for the silhouette of a spherical nest.

When searching a tree from the ground, be methodical. If you can (for example with a big oak), follow each branch with your binoculars. Check the tree out from all sides. If you can, try to get a side view of the canopy from a distance, perhaps by climbing up a valley side. When nests are high up in the canopies of big trees, they are often only a couple of feet (60 cm) inside the canopy, and a breeze might

Asian hornet control

BOX 4. Bait 'shadow'

"I got it down to a consistent minute and now I've moved the bait station in the same direction I'm getting three minutes! What is going on?"

One situation noticed in Jersey was the timings going haywire if you are too close to a nest. This did not happen for a low nest, but for nests high up in trees.

Suppose you move your bait station in the direction of flight, but you move it to a place directly under the tree with the nest in. You don't know it yet, but the nest is 21 m above your head! In this situation the hornet may not be able to climb vertically carrying a heavy load of bait – she will fly at a shallower angle and have a longer flight time, plus she may not recognise the area under the tree, because she normally flies out of the canopy and off to forage, staying high for a while. So it can take her longer to get to and from the nest when she is directly under it, than it did 100 m back along the route you have just taken.

The bait station further from the tree (blue) gives a more direct and faster route for the Asian hornet than the bait station beneath the tree (red)

BOX 5. Attaching a fluff
Text and photos by Nigel Errington

I have found the following products work best, after rigorous trial and error testing:

Gulff UV resin (the brighter the better)
Orvis Egg Yarn
Uni Big Fly thread, 400 denier

Preparation

Tie a slip loop (I use an improved clinch knot) knot in a 100-mm length of cotton. The length of cotton is unimportant as long as you have sufficient to trim shorter later. Then select a suitable 25-mm length of yarn (about as thick as a match-stick).

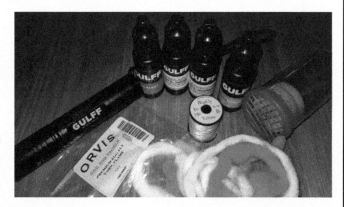

Pass this piece through the loop and pull tight. This will be the starting point; when attaching to the hornet you will reduce the overall length to about 25 mm but I have found some larger hornets are capable of carrying a larger load.

Method

Catch a hornet in a queen-catcher (the plunger type). If you snip out a piece of the plastic grid, you will have a larger rectangular hole to work through: this gives a better holding point for the thorax and allows more work space. When you have your hornet you need to entice it to get its top side to the grid and gently hold it in position so its thorax is aligned with the rectangle. (Yes, patience is required.)

Apply a little UV resin, only a micro dot, about a millimetre in diameter, to a segment two or three away from the tip of the tail (on the top side). Set with the UV torch.

Asian hornet control

Once this dot has set you need to attach the end of the prepared cotton and yarn. Cut to length and with the assistance of a queen catcher holder (a brave second person) add a 1-mm drop of resin to the end of the cotton, and using the tweezers apply and hold this end on the already applied first blob and set in position, ensuring that the cotton and fluff point to the rear. Now trim the yarn to about 10 mm either side. Remember this can be reduced if the hornet struggles with this weight.

Slowly release your hornet and use the tweezers to help the yarn pass through the holes in the queen catcher. Now let your hornet loose in a small flight tent if you have one and allow it to settle down; you may also find that the hornet settles back on the bait station to feed. When the hornet flies off it will remain visible against difficult backgrounds and so will be visible for much longer.

part the leaves just enough for you to see it. Sometimes the time of day makes a difference — light coming from a different angle can help or hinder.

Apart from picking out that beige ball up in the canopy, nests can also be discovered (or the tree confirmed) by spotting 'traffic' — the fast and direct movement of hornets in and out of the canopy (roof space, hedge, brambles). Through binoculars, tree canopies seem smothered in insects (they are), but most are bumbling around, dipping in and out, flying here and there, and they are all different sizes. The traffic going in and out of a nest is purposeful and unswerving, and the hornets themselves are big and distinctive compared to other insects around. Commonly, the traffic goes into the canopy from above, even vertically above, when it can resemble (with a little imagination) some kind of Roman candle firework!

Most nests are high up in tall trees but some nests are built at much lower levels, in hedges, brambles, wall cavities; a very small percentage are even found underground. Of the 36 secondary nests found in Jersey in 2018, 23 were found in trees (roughly 64%), 8 were found in building 'voids' (e.g. attics or behind soffit boards) or on buildings, 3 were found in hedges and 2 in bramble patches.

Low locations, especially hedges and bramble patches are very dangerous. When the nest is high up, it poses no danger to the hornet hunter, but when it is low down, it can be extremely dangerous. This is because Asian hornets will aggressively defend their nests. Always be on the lookout for individual hornets — if you are investigated or bothered by an individual hornet, get out IMMEDIATELY. If you accidentally get within a few metres of a nest ('about 5 m', de Haro et al. 2010: so double this at least to be safe), you may be attacked. When this has happened to people, a single sting acts as a signal to stimulate more hornets to attack. People have been hospitalised for multiple stings, and this can even prove fatal if you are allergic or have a condition that makes you susceptible to the effects of the venom (for example some heart conditions).

Once you have found the nest, make sure you can find it again — perhaps even leave something at the base of the tree — and contact whoever is going to destroy it (remember that this section portrays a scenario in which permits have been granted, which is likely to be after Asian hornets have established). Hopefully you will be working with the NBU, but if the NBU no longer has this role, it may be a local pest controller who will destroy the nest, as happens in France.

UNDER NO CIRCUMSTANCES TRY TO DESTROY THE NEST YOURSELF

What Three Words

What Three Words is a phone app for individually naming every square metre on the surface of the planet. If it is difficult to describe a location (of a sighting, a bait station, a nest), you can use the satellite imagery on the app to pinpoint the location, and the app will give you a unique three-word code for that square metre. You can then keep that code to record the position of something, or give that code to others and they can find that point again.

Secondary nests. ***Top left****: nest in hedge — very dangerous and difficult to see.* ***Top right****: glimpse of nest in a tall tree. Often there are only certain spots where the nest can be seen from the ground.* ***Bottom left****: nest in sycamore with hornets on the surface.* ***Bottom right****: a much easier nest to spot, high up in a pine tree, but away from foliage.*
Top right photo by John de Carteret, the rest by Judy Collins

Zooming in on a nest in a sycamore. Photos by John de Carteret

Asian hornet control

Zooming in on a nest in a sweet chestnut. Photos by Judy Collins

Radio Telemetry

Sarah Bunker and Peter Kennedy

Radio telemetry has been used by biologists for many years to locate and track individual animals fitted with radio-tags (transmitters). You are probably already familiar with wildlife programmes in which large mammals, birds or reptiles are fitted with collars or glued-on devices used to indicate their whereabouts. This is fairly straightforward when you are dealing with, say, a wolf where the radio-tag weighs a fraction of the tagged individual, but when you want to track insects, you start to rub up against the limits of miniaturisation of the hardware.

The radio-tag generates an individual radio signal, powered by a battery and sent out along a thin aerial. The smaller the radio-tag, the smaller the battery, and the weaker and more short-lived the signal is forced to be.

In order to follow the insect, the operator uses a directional antenna attached to a hand-held receiver, which picks up the signal from the tagged insect. The signal is transformed such that its strength is discernible on a visual display as well as audible over built-in speakers. Although signal strength is not a reliable indicator of distance, it is an accurate indicator of the direction the tagged insect is from the operator.

Number 32 is fully awake now. She has been gently restrained by a pliable wire which crosses her petiole (waist). The radio transmitter has been attached (black object below abdomen) by tying with cotton thread around her waist; you can see part of the aerial which will trail behind her

Asian hornet control

Dr Jess Knapp uses the antenna and receiver to track a tagged hornet to its nest. Not all nests are found in the wilds!

Prof. Juliet Osborne's team at the University of Exeter's Environment and Sustainability Institute in Penryn, Cornwall, have a long history of studying the foraging behaviour and flight paths of social bees, and Defra were interested in how these could be adapted to track hornets.

The first problem they met was finding transmitters small enough to be carried by the hornet. Because Asian hornets can carry quite large prey back to their nest, they are at least built to lift and, by experimenting, the team found that an Asian hornet could successfully fly with a tag up to an astonishing 80% of its body weight, including a 10-20-cm thin wire aerial which trails behind the tag. This still meant that they had to find workers weighing at least 278 mg or 350 mg (depending on the weight of the tag used), which can be a problem, especially before August.

The tags are attached with cotton thread around the waist of the hornet. A tricky process that is best achieved by cooling the insect in a tube buried under crushed ice until it has become torpid, and then restraining it briefly while the transmitter is tied on. This technique is not for the faint hearted, because the hornet begins to 'wake up' in about 2 minutes, and you want to know it is secure. Dr Peter Kennedy has developed a clever acrylic plate with a restraining loop, that holds the hornet in place while you tie the tag on, yet allows you to easily release the hornet with the tag in place. Once released, the tag hangs below the hornet but is sufficiently mobile to allow her to fly and walk. Although the procedure for attaching the tag is straightforward, you do need a balance that is reliable down to at least 0.01 g; a good hand-lens or basic dissecting microscope are also recommended to check that knots are secure. After the tag has been attached, the hornet is placed in a small tent with food so that she can recover from the cooling process and get used to the tag. Once she is showing good signs of recovery, she can be released.

The set-up for tracking the hornet consists of a directional antenna, which superficially looks like an old-fashioned TV aerial: it is excellent in picking up the direction of the radio-tag. This is attached to a receiver, a box which is carried via a neck/shoulder strap. The radio-tag on the hornet produces short 'pips' of radio signal, at regular intervals, that are converted to audible 'pips' by the receiver. Once the hornet has flown off, the operator follows by sweeping the antenna in an arc to determine the direction with the strongest signal, and sets off in pursuit. A clear advantage is that when obstacles are in the way (walls, lakes, private property with no access granted), the operator can move around these and re-acquire the signal from the other side.

In 2017, Dr Kennedy tracked hornets in both the area around INRA Bordeaux-Aquitaine and on the Channel Island of Jersey (in both urban and rural settings). The amount of time taken to actively track a tagged hornet to an unknown nest ranged between 45 and 133 minutes (mean = 92 minutes). Nests were found between 195 m and 1.3 km from release points. The chase can be slowed down by the hornet stopping to rest, feed on nectar-rich flowers, or to attempt to remove the tag.

Depending on the size and model of radio-tag used, those suitable for insect tracking have ranges between 375 m to 1 km in flat open landscapes and batteries guaranteed for at least 4-12 days. Usually the hornet will return to its nest, so by patiently following, it has proved to be a very successful technique. In practice, the method works best with two people, one person listening to the receiver and the other watching the terrain, getting permissions to go onto new land and keeping a look out for the nest. Listening to the strength and quality of a signal is easier when other noise, such as that from other excited hornet hunters, can be kept to a minimum.

The equipment itself can operate in all conditions hornets do, and although the tags are not cheap (£140 + VAT each), they can theoretically be retrieved for re-use.

This technique has been used successfully in France, Jersey and the UK, and provides the NBU with a useful alternative when nests have not been found fast enough by triangulation methods. Further miniaturisation of the radio-tags would help extend the usefulness of this approach.

Other methods of finding nests

Apart from tracking using flight lines and timings, or radio telemetry using transmitting tags, here are some other approaches which have been taken.

Using drones

Drones, or UAVs (unmanned aerial vehicles), would seem a natural fit to finding big nests in trees that may be hidden by foliage from below. Unfortunately, they are usually hidden by foliage from above, too. OK, so how about fixing up a UAV with an infra-red video camera? As we have seen, Asian hornets heat their nests to around 30 °C (although they are well insulated by the nest walls, which are paper pockets filled with air, they may still be quite a bit warmer than the background), so perhaps this heat signature could be picked out against a cold background: perhaps at dawn? Well, this technique has been tried, but hasn't been very successful.

In Jersey, when the first secondary nest was found, the Jersey States Fire and Rescue Service tested their UAV with infra-red video equipment on board. It was flown close to the nest in the cool of the early morning to see if there was a useful heat signal, but sadly there wasn't. While the drone was near the nest it was investigated by the hornets and collisions with the rotor blades could be heard from the ground. The hornets could be seen attacking the drone, and it was hoped that this could be used for nest detection. However, when the drone was used at another nest, no attack response was triggered.

The National Bee Unit has also tried a UAV with an infra-red camera at the Brockenhurst nest, but it was unsuccessful, partly because the tree canopy was too warm to provide good discrimination, and possibly because the nest had suffered recent damage and so was unable to maintain its heat.

A different type of approach has been put forward by Reynaud & Guérin-Lassous (2016), who propose using a small swarm of UAVs to follow a hornet fitted with a lightweight but very visible marker. They assume that the insect will fly in a straight line back to the nest, which is not the experience of trackers so far (and hornets fitted with markers or tags sometimes stop to rest and/or attempt to remove the object). It will be interesting to see how this research develops, because it does have great potential.

Harmonic radar

Harmonic radars have been used to track insects for many years now. The radar sends out a wave at a fixed frequency and receives a reflected signal from a metal tag attached to the insect. Previously, this type of tracking has been most successful in flat, open spaces, where the signal was clear; in hilly and woody landscapes it was difficult to make out the signal amongst the noise. In 2016, some Italian scientists (Milanesio et al. 2016) reconfigured the setup, using a small marine radar and advanced processing techniques to pinpoint Asian hornets in particular, in order to find their nests. In 2017, the same team experimented intensely and ended up with a detection range of 150 m. Their idea was to slow the advance of the Asian hornets in the northwest of Italy by setting up the radar in apiaries under attack, and using it to track them to their nests, followed by nest destruction.

Insect energy harvesting

Another interesting piece of research, from Bangor University, has looked at harvesting energy from an insect in flight and converting it to electricity which could power a transmitter (e.g. a radio-tag) (Shearwood et al. 2018). The system uses the flexing of a tiny strip of material which vibrates as the insect flies to generate electricity (piezoelectric generation). This sort of technology could help the miniaturisation of radio-tags which are not yet light enough to work with the smaller Asian hornets found earlier in the year.

Asian hornet colony destruction

At the time of writing (2019), the destruction of any Asian hornet nest (primary or secondary) should be done by the National Bee Unit (NBU), as part of their eradication programme.

The NBU employs professional pest controllers who have been trained in dealing with Asian hornets. They have specialist equipment: a lance or cherry-picker to reach high nests, and licensed insecticide. They also remove the nest the following day to prevent insecticides from harming local wildlife.

Destroying primary nests

In France, destruction of embryo nests (the initial nest containing only the queen) is often done by beekeepers. However, once the workers start emerging, it should be treated as being as dangerous as a secondary nest, and the professionals called. Wearing a bee suit, many French beekeepers employ various tactics to remove the entire nest with the queen still inside, including a simple jar-and-card technique (AAAFAc 2016) (and finishing them off in the freezer), or a clever clam-shell device made by taping two small sieves or strainers to the jaws of a litter-picker, for a long-armed approach (AAAFAd 2015). This device can also be used to catch queens on their favourite camellia flowers (AAAFAb 2016)

Destroying secondary nests

The nest should be destroyed at dusk or during the night: Asian hornets are generally all home by dusk and do not forage again until first light. If the nest is removed during the day, the foragers away from the nest survive, and will start to build a replacement nest at the site of the old nest the next day. If a poisoned nest is left in situ, foragers are poisoned on their return.

The hornet suits being used in Jersey were thick and heavy, made from tightly woven fabric, with a flexible plastic face visor. When I tried one on it was pretty awful — hot, steamy and cumbersome — and that's without actually trying to do anything practical while wearing it. Steve Bright from BBwear recently went to France to visit Robert Moon, an Englishman who has taken French pest control qualifications and works in Bessais-le-Fromental in central France. There they tried out BBwear's 5-mm mesh suit during operations to destroy and remove Asian hornet nests, with excellent results.

Equipment needed for nest destruction using a lance (to right, in sections), a duster (at back) and CO_2 to propel the insecticide (not shown).
Photo by Sue Baxter

The only insecticide licensed for use with Asian hornets in the UK until the end of 2018 was a product called Ficam D, which has bendiocarb as its active ingredient. However, at going to press, this insecticide is no longer allowed to be used outdoors in the UK, and is no longer approved at all for use in the rest of the EU. The NBU is looking at alternatives, but may try to obtain a dispensation to use it in the UK considering that they are able to enforce stringent protocols for use; in its favour is its fast action. Insecticides often come as fine powers or liquid suspensions. They have to be delivered into the nest as broadly as possible in order to coat as many insects as possible, including the queen. If the nest is low down, then a duster or sprayer can be used, which uses a pump-action pressuriser. Powder is delivered via a hose and hollow spike, which is jabbed through the nest wall. The reach can be extended using a telescopic lance. For very high jobs, either a cherry-picker needs to be employed and someone goes up with a duster, or a specialist really long lance is used. These long lances can be up to 30 m and the insecticide is packed into a chamber near the top. CO_2 is used to fire the powder into the nest.

Although the pesticide is supposed to remain in the nest, it often blows out of the nest through the entrance or breaches made by the spike, or the force of the CO_2 can blow away the nest wall. It can also be carried at least 100 m by coated insects if the operation is carried out during the day. The NBU spread a sheet below nests they destroy to collect any dead hornets which drop out of the nest.

In France, the lance technique is commonly used, although cypermethrin and permethrin seem to be the insecticides of choice. Sulphur dioxide was also used during a short period of time when it was deregulated specifically to deal with Asian hornets. Turchi & Derijard (2018) argue that although sulphur dioxide is dangerous for a user in the event of a leak in an enclosed space, it is much less harmful to the environment than permethrin, and cheaper. Although nests treated with insecticides should be removed from the environment, in practice nests are rarely removed from high locations such as treetops after poisoning in France, but are usually removed from sheds, houses etc.

Turchi & Derijard (2018) also note that other biocides such as pyrethrum or diatomaceous compounds could have less of an impact on the environment than permethrin, so perhaps should be used

Destroying a nest in a hedge. Photo by Judy Collins

instead, once they have been fully assessed. UAVs ('drones') have been used to deliver poison into nests (see video: Chibrac, 2013).

Another French delivery option for insecticide is a patented modified paintball gun which fires small balls (0.68 inches) into the nest. The balls contain fairly standard insecticides [pyrethroids: the PILP and DIPTER BALLS are registered and patented, and you have to have a 'certibiocide' (biocide handling certificate) to be able to buy them]. However, the balls that miss the target will be delivering insecticides into the environment, as well as those that lodge in the nest.

Urgent need for non-insecticide nest destruction

Because of the problem of insecticides affecting wildlife, other ideas have been put forward for nest destruction, perhaps using fungal spores (see 'Biological control'), CO_2, water or steam, and perhaps delivered using a UAV. Delivery of traditional insecticides using a UAV has been achieved, but it is not straightforward, especially when the nest is well inside the canopy; it can be difficult to negotiate branches with a UAV big enough to carry a decent payload, and a long delivery tube is required. Other techniques tried include hoovering the insects from a nest: certainly an option for low nests. In Mallorca, Wildlife Rangers with responsibilities for nature reserves remove nests in these sensitive locations without using any insecticides at all. Already trained in tree climbing because of their wildlife duties, they climb up to nests, plug the entrance, free the nest from the tree, bag it and take it down to be destroyed (Peter Kennedy, pers. comm.). This is an extremely tricky operation involving a stealthy approach at night, wearing a hornet suit: you have to be extremely dedicated to do that!

Left: *insecticide drifts away after being fired into a nest. Photo by Gerry Stuart.* Right: *this hornet, covered in insecticde, dropped out of the sky more than 100 m from the nest*

Asian hornet control

There is an urgent need for techniques that will kill off all the adults in a nest at night, but leave no lasting harmful residue, so that nests can be left in situ. All you tinkerers and inventors, get your thinking caps on.

This page. **Top**: *Asian hornets busy rebuilding a nest that was destroyed during the day: these would have been out foraging at the time the nest was destroyed. Photo by John de Carteret.* **Below left**: *the paintball gun reconfigured to fire insecticide-filled balls (www.frelons.com).* **Below right**: *Bob Hogge in a hornet suit with chunks of nest. Photo by Gerry Stuart*

Using a very tall cherry-picker to remove a nest in an oak tree.
Photo by Gerry Stuart

Trapping

Context

Insects are in trouble, and we are just waking up to this fact. In the UK, where insects are comparatively well studied, there seem to be the largest documented declines across groups (60% of species). The drivers for worldwide insect decline appear to be (in order): (1) habitat loss and conversion to intensive agriculture and urbanisation, (2) pollution, mainly that by synthetic pesticides and fertilisers, (3) biological factors, including pathogens and introduced species, and (4) climate change (Sánchez-Bayo & Wyckhuys 2019). The authors looked at 73 historical reports of insect decline (mainly in developed countries), and found dramatic rates of decline, which could lead to the extinction of 40% of the world's insect species over the next few decades. This loss of insect biodiversity could be disastrous for the Earth's ecosystems, because insects are indispensible in food webs, are part of many other organisms' life histories (for example, pollination), and help in many great cycles of elements.

Bearing this in mind, we don't want to practice indiscriminate trapping of insects in our single-minded focus of getting rid of Asian hornets. We need to find the best ways of removing Asian hornets without endangering the others. It's a difficult path to tread, and any action needs careful consideration.

Monitoring traps

At time of writing, the UK is getting regular incursions from continental Asian hornets, but there is no evidence that they are established here. It is essential to monitor for Asian hornets so that if any are detected they can be dealt with swiftly by the NBU before any gynes (queens) are released from the nest, which could start new colonies. This monitoring can be passive: keeping an eye on natural nectar sources (e.g. camellia flowers), sap sources (e.g. wounded trees), protein sources (e.g. honey bee hives, food waste processing plants, meat and fish outlets), or it can be active: by actually providing something attractive on purpose ('bait'). Monitoring can be carried out at any time of year Asian hornets are active (March/April to late November/early December: we don't know for sure the timings yet): from cherry blossom to first heavy frosts.

When does a monitoring trap become a killing trap?

Any trap that is neglected will turn into a killing trap. In fact, many insects cannot survive even a short period (a few hours) in a plastic trap, especially when there is liquid bait to drown in: heat, exhaustion, drowning, getting eaten by a predator in the trap, and getting covered in sticky material are all factors leading to insect deaths in traps.

It is also important to note that just because a trap has exit holes, it doesn't mean that non-target insects are able to escape through them: they may not find the holes, or recognise them as escapes. Monitoring traps should have any attractant (bait) soaked into a material (e.g. sponge), or placed beneath a mesh, so that insects don't drown and, importantly, should be checked regularly (at least once a day; better still, twice a day) so that non-target insects can be released.

Monitoring without trapping

Because Asian hornets will come to dishes of attractant to feed, do you actually need to trap them to discover whether they are in the vicinity? Having a bait station like those used when tracking Asian hornets means that you can monitor without trapping, but you will need to put it somewhere that you observe frequently during the day: perhaps outside an office or kitchen window. If you can organise a roof above it to stop it filling with rain and diluting the attractant, it can be left in place. A Véto-Pharma-type trap with Suterra (or the Véto-Pharma attractant supplied with the trap mixed with some beer to put off bees) absorbed into material (e.g. a sponge, tissue paper) and the entrance funnels removed makes a good non-killing monitoring trap that insects can get in and out of (although the yellow plastic of the beaker can make it harder to identify insects through). Another avenue to explore in terms of monitoring is a camera trap in conjunction with a bait station; some camera traps are sensitive enough to photograph Asian hornets visiting the station.

Kill trapping

Spring trapping is done to catch and kill Asian hornet queens early in the year before they create a proper colony.

Summer/early autumn trapping is done to relieve predation pressure on honey bee colonies: it is the workers that are trapped and killed.

Asian hornet control

The NBU monitoring trap, made from a 2-litre PET bottle. It has a mesh floor and sweet liquid bait is poured in to just below this mesh. The mesh prevents insects from drowning, and the roof stops the trap from filling up with rain, as well as disguising the entrance hole to prevent a hornet from finding its way out. A link to the instructions is on p 138. Note that there are no escapes for non-target insects.

Late autumn/ winter trapping is done to catch and kill the sexuals (males or gynes) before they mate, or gynes (queens) before they hibernate.

I'll go into more details about these different types of trapping later.

Trapping is contentious

At present, trapping of Asian hornets is done using food traps (sweet or protein baits), as opposed to pheromone traps (non-food-based scents specific solely to Asian hornets). The problem with food traps is that they are not selective: they attract all sorts of insects. The non-target insects caught in a trap are sometimes called the 'by-catch'. If a pheromone trap becomes available, it may solve the problem by avoiding catching other insects, but for now the by-catch is a serious problem. Out of all these different types of trapping, spring trapping for queens is the most contentious of all.

In broad strokes, in France there are two camps when it comes to spring trapping. On the one side, there are those (mainly ecologists/biologists) who think that traps used to catch Asian hornet queens in the spring are simply not selective enough to justify their use. Because they are unselective, the number of non-target insects caught is unacceptably high for the number of Asian hornets caught. They also believe that spring trapping catches too few Asian hornet queens to have any significant impact on the number of nests in a region.

On the other side, there are those (mainly beekeepers) who think that spring trapping of queens is effective in reducing numbers of Asian hornet colonies.

The effect of trapping on non-target insects: the data from France

There have been several studies on the effects of Asian hornet trapping on other insects (entomofauna); a few non-insects get trapped, too, but thankfully very few. Most studies have looked at spring trapping, when Asian hornet foundresses (the fertilised queens that have hibernated over winter and emerge to start new colonies) can be trapped; there is much less information on late summer/autumn apiary trapping of workers, and even less on trapping of males and gynes (the sexual generation) in the autumn.

In 2007, Jacques Blot, a study leader for the ADAAQ (Beekeeping Development Association in Aquitaine), wrote a technical sheet about spring trapping of Asian hornet foundresses. At that time more than six regions had Asian hornets and he promoted a massive trapping campaign by beekeepers and the public, to be initiated when captive *Vespa velutina* emerged from their overwinter hibernation (to indicate the right time to set traps). To avoid by-catch, he proposed a 7-mm entrance hole and 5.5-mm exit holes to allow non-target insects to escape. Traps were to be inspected daily or weekly, with non-target insects released, and data to be collected.

In 2009, Dauphin & Thomas conducted an experiment to survey the trapping efficiency of wasp traps (the ones half-filled with liquid, which kill everything), baited with beer, white wine and sugar syrup. Their survey was during the summer (June, July, August), and they found that the traps caught 87% flies: 1000 a week per trap, 10% butterflies and moths, 2% beetles and smaller percentages of other insects, with Asian hornets coming in at 0.57%.

In 2010, Haxaire & Villemant tested Blot traps (unfortunately, they didn't follow his original trap plans, which had a 7-mm entrance: instead the neck of the bottle was left open, making an entrance of perhaps 20-25 mm: they don't say how big). They used a bait of lager, cane sugar and rum and tested the traps from the end of March to mid-May. In this experiment, 1,200 insects were caught, mainly ants, and then flies, butterflies (mainly speckled woods) and beetles, various other insects…and eight Asian hornets, in 90 traps over 8 weeks — a catch rate of 0.01 foundresses per trap per week, with Asian hornets representing 0.67% of the insects caught.

In 2012, Monceau et al. looked again at spring trapping, using two types of traps: one a killing trap with no escape holes and liquid bait, and another with a 25-mm entrance and 5-mm and 5.5-mm escapes, with bait absorbed into a sponge (probably the most selective trap at the time; it has since been improved by reducing the size of the entrance hole). They trapped in two locations, from the end of February to the beginning of May. The overall yield of their trapping was 0.71 foundresses per trap per week, with Asian hornets representing 1.7% of the insects and others caught. This difference in numbers caught compared with the previous study was put down to differences in population numbers and perhaps weather conditions. In their conclusion, they say that foundress trapping is 'a drop in the invasiveness ocean', but they also acknowledge the trade-off between loss of entomofauna in traps, and loss of entomofauna from worker Asian hornets later in the season. They suggest more specific traps, and await the production of specific pheromone traps.

Does spring trapping reduce the number of Asian hornet colonies?

In 2013, Rome et al. surveyed a couple of areas where spring trapping had been conducted by local associations. In the Pays-de-la-Loire region in 2008 a first individual was found south of the Vendée, and in 2009 a limited number of selective beer traps were set up within 30 km of the first sighting. No foundresses were trapped, and 12 nests were later found. In 2010, 400 traps were set up around apiaries, with a big public education effort. Six foundresses were trapped and 195 nests were later found. In 2011, 12 foundresses were caught and 485 nests were found that year.

Elsewhere, a spring trapping operation began in 2007 in the Bordeaux region. It was extremely intensive, with up to 15,000 traps in 550 square kilometres (around 27 traps per square kilometre!) by 2009. Several thousand foundresses were caught each year, but the number of nests found remained stable at about 300 each year.

Similar experiments have been conducted at municipal and departmental scales without any effect on nest densities, comparing years with and without trapping. (Dordogne in 2007, Test-de-Buch 2010, Andernos-les-Bains, 2011, etc.). Conversely, no massive trapping campaign was carried out in the Lot-et-Garonne between 2007 and 2009, yet the numbers of nests detected halved between 2007 and 2008, from 609 to 267, and slightly recovered in 2009.

In Morbihan, the following results were found:

2011	5 nests found	
2012	63 nests found	
2013	235 nests found	
2014	1150 nests found	
2015	2918 nests found	co-ordinated spring trapping and nest destruction begun
2016	5062 nests found	co-ordinated spring trapping and nest destruction
2017	3089 nests found	co-ordinated spring trapping and nest destruction
2018	4178 nests found	co-ordinated spring trapping and nest destruction

What is going on here? In 2015, 2016 and 2018 the numbers of nests increased, despite large-scale co-ordinated spring trapping and destruction of nests. Spring trapping should reduce the number of nests in the same year, if it works. Nest destruction, if it happens before the sexuals emerge, will have an effect on the numbers of nests the following year. In 2017, there was a drop in nest numbers: was this related to winter or spring conditions (Rome et al. 2013), spring trapping in 2017, nest destruction in 2016, or some other factor?

The organisations that gathered this data (FDGDON du Morbihan and FREDON Bretagne: see 'Resources' for links) have joined forces with ITSAP (Institut technique et scientifique de l'abeille et de la pollinisation) from 2016, to conduct a multiple-year data collection exercise, which hopefully will clarify what is going on.

Usurpation — natural population control?

Several authors talk about usurpation being a natural mechanism for population control, which the killing of hibernating queens or foundresses disrupts. In New Zealand, when the widespread killing of hibernating wasp queens (*Vespa germanica* wasps were introduced into New Zealand in the 1950s) was encouraged by a bounty paid per head, the number of colonies later that year actually increased (Martin 2017). It is thought that this unexpected result was due to the removal of competition between queens. M. E. Archer, who wrote the chapter on population dynamics in Edwards (1980) calculated the mortality rates of queens at various stages, and found a 98% loss of queens over the winter, presumably due to weather, parasitism (including fungi) and predation, although heavy rain and low temperatures did not seem to kill hibernating queens, and winter temperature did not influence wasp numbers the following year. A further 92% loss occurred in the spring due to bad weather, migration, predation, competition for nest sites and quality of queens (Edwards, 1980). Out of all the factors that could limit the wasp population, competition between queens seems the most significant. In wasps, the usurpation of primary nests is very common and widespread, with at least 30% of colonies having undergone usurpation and an average of 12 queen changes in a nest (Archer in Edwards, 1980).

Usurpation is often cited to discredit the trapping of Asian hornet queens in the spring for population control (Monceau & Thiéry 2017; they quote Haxaire & Villement 2010). The argument goes that if you don't try to trap the queens, then many of them will kill each other anyway. Conversely, if you trap queens, then the ones that don't get trapped will be more successful because they won't be challenged, and won't get injured. Martin (2017) notes that in Japan, in a year when there are many embryo nests and much usurpation, few large colonies appear later in the year (presumably he is talking about hornets here, but he does not give a species).

The only method that did decrease the density of wasp colonies in New Zealand was an almost complete destruction of nests: killing queens in the autumn or spring was completely ineffective.

UNAF on trapping

The views of the National Union of French Apiculture (UNAF: Union Nationale de Apiculture Francaise) illustrate the other side of the argument. In 2017, they wrote a small article about spring trapping (UNAF, 2017) as a reply to two environmental groups who were against spring trapping [France Nature Environnement (a federation of French nature and environmental protection organisations), and OPIE (Office pour l'Information Eco-entomologique, a French governmental organisation devoted to entomology, especially applied entomology)]. UNAF argues that done at the right time and in the right places, spring trapping is an effective means of control, which helps prevent the damage caused to local wildlife and bees by *Vespa velutina* later in the year. The evidence they give for the effectiveness of trapping is as follows:

In Trélissac, Dordogne, where trapping was well conducted by the city, the number of nests was significantly reduced from an average of 40 nests in previous years to between three and five nests in years when spring trapping was practised.

In Morbihan, where spring trapping was organised at the departmental level, they say that the number of hornet nests has significantly decreased where trapping has been well conducted (but see previous data), with the most colonised commune remaining the one that hardly traps (Lorient). A big problem with this data is that increased spring trapping and nest destruction go hand-in-hand: when spring trapping is stepped up, so too is effort in nest destruction. Even if you could prove that intervention was reducing numbers of colonies, you still could not say whether it was down to spring trapping or nest destruction.

Where others stand in France

Doing this research has been quite frustrating in that the ecologists, who on the whole oppose spring trapping (but not necessarily late summer apiary trapping when predation pressure in apiaries is high, or late autumn trapping to catch gynes) show their data clearly and openly: the organisations on the ground, like UNAF, FREDONs, FDGDONs, mainly give anecdotal results and if they have data, they do not seem willing to publish it. And on the part of the ecologists, it would be good to investigate numbers of non-target insects caught in traps with smaller entrances: this would really help in understanding the parameters of trapping.

More data is needed to disentangle the effects of spring trapping, efforts made to find and destroy nests, winter and spring weather, and natural population dynamics on numbers of nests from year to year.

Some big questions that need answering include: Are their fewer nests this year, or haven't you found them? Are there more nests this year, or have you been looking harder? Also, do numbers of nests reduce naturally after a few years? And what about the interactions of populations when one area uses trapping and the adjacent area does not? Do you get an influx from marginal populations?

The MNHN (Museum National d'Histoire Naturelle: National Museum of Natural History) is very much against spring trapping, because of its effects on other insects. INRA (L'Institute national de la recherche agronomique: National Institute for Agricultural Research) feels the same way.

Grassroots organisations, such as the AAAFA (Association Action Anti Frelon Asiatique: Anti-Asian Hornet Action Association) are clearly for spring trapping and argue that it is the most effective way to lower Asian hornet numbers.

Conclusions from French spring trapping data

At present there is not enough evidence to support spring trapping of Asian hornets in France: it seems to make no difference to numbers of nests later in the year, and a catch rate of only 1-2% Asian hornets is very ineffective: the practice could endanger populations of non-target insects in entire regions (Rojas-Nossa 2018).

In the UK, of course, we are monitoring for Asian hornets from March to December, and will be most likely to see or trap an Asian hornet worker in September or October when their numbers will be highest if a colony has set up.

We are waiting to see the results of the ITSAP study.

Trap practicalities

Can traps of any sort be justified for trapping Asian hornets in the UK at the moment?

The NBU strongly encourages monitoring traps, especially along the south coast of England, and provides directions for making them. Certainly, there are places that should be monitored, like areas where nests have been found, and in all the sentinel locations, which include ports and airports. Last year, two Asian hornets were found in traps (Fowey and Liskeard), the former leading to the discovery of two nests and the latter most likely coming from Fowey.

The predicament in the UK at the moment is that we need to know where Asian hornets are so that we can eradicate them; yet, on the other hand, because they have not established, they are rare and patchy in distribution. Monitoring traps can certainly be part of the mix in finding them, but they need to be used with respect. If you are a methodical sort who can be depended on to check monitoring traps daily, then go ahead and set some up. If you intend to check a trap, but a week goes by and you haven't, then remove it.

If you are a beekeeper, then observe your hives for hawking hornets; if you are a flower-lover, then observe plants that attract a lot of insects. We need to monitor – that much is clear; an unobserved bait station is useless, and a neglected monitoring trap kills insects unnecessarily.

Diversion traps

The aim of diversion traps is to attract and trap hornets a short distance away from the apiary, hopefully providing them with a food source easier than catching honey bees. The problem with this is that you are more likely to catch non-target insects. If the colonies are being attacked, it is better to have killing traps right in amongst the hives where non-target insects are less likely to be caught.

Trap design

In areas where Asian hornets have established, there are four types of trap generally being used. The first is the manufactured wasp-killing trap. Examples are the dome-type where the insects enter through a wide funnel from beneath (e.g. Pest Trappa), the beaker-type where the insect goes in through a funnel at the side, or top, which has a roof over it (e.g. Véto-Pharma, which can be used with or without the trapping entrance, if you want to use it as a bait station); and plastic sets of roof plus funnel, which are screwed onto your own jam jar (e.g. Vaso Trap made by Tap Trap). Manufactured traps are sturdy but relatively expensive, costing around £4-£10 per trap. They usually involve yellow plastic, as yellow is attractive to wasps, and are generally designed to be hung from a tree, but can be placed in a holder (e.g. brackets, small bucket etc.). They were originally designed for killing wasps, and in this mode they are filled about half full with a liquid attractant (e.g. Suttera, jam and water or half-and-half honey and water, fermented). They are unselective: anything attracted to sweet bait will crawl in through the funnel-shaped opening and will not be able to get out. Eventually, any insects caught in the trap will get exhausted and fall into the liquid and drown. The bodies just keep piling up, which is the aim. These could only be condoned in an apiary suffering extremely high Asian hornet predation (more than 10 hornets hawking each hive).

Véto-Pharma trap being used as a monitoring trap. It has quite a few wasps in it: some have died, probably from exhaustion. Photo: Judy Collins

Roof of Véto-Pharma trap peeled back to show lid with entrance funnels: the lid can be removed to make a bait station.

The second type is the home-made plastic bottle trap. These have been evolving from mimicking the ordinary wasp trap to becoming relatively more selective. The basic pattern involves a plastic bottle with the top third cut off and inverted to form a funnel trap. After that, it's all about not killing the insects that get inside so that you can selectively kill *Vespa velutina*. The modifications include: (1) making the entrance hole too small for native hornets [7mm according to JP33, 8 mm according to Blot & Debellescize (2017)], but allowing Asian hornets in; (2) avoiding drowning insects by using attractant soaked into a sponge or paper, or placing a screen between the insect and the bait (as in the NBU trap); (3) allowing insects smaller than the Asian hornet to escape [5.5 mm according to JP33 and Blot & Debellescize (2017)]; (4) having a way to take the trap apart to release insects; and (5) having a roof or cap to stop the bottle filling with rain and so drowning the occupants. The colour of the roof is important if the funnel is directly beneath it. The underside should be black so that trapped Asian hornets do not see the open funnel they came in through as an exit: they associate escape with light.

*Home-made monitoring traps. Photo **above right** shows how a plastic bottle trap has been put together. Caps (from the same types of bottle) are threaded onto plastic-coated garden wire: these are spacers which provide an 8-mm entrance gap between the top of the funnel and the roof. **Above left**: close-up of 'escape hole' for smaller insects, 5.5 mm tall. The ramp leads up from pieces of sponge soaked in Suterra.* **Insects are unable to find the escape hole and instead get trapped at the top, between the funnel and the bottle wall, therefore this type of trap is not recommended.**
***Below**. The author has been experimenting with an old bird feeder with 5.5-mm wire mesh. Again, an inverted plastic bottle funnel has been used, with a 7-mm hole drilled in the cap. She intends making a metal gauze side funnel which narrows to 7 mm, instead of the plastic funnel with drilled cap, which*

might be an obstacle for Asian hornets. Smaller insects can come and go as they please: none trapped so far. Top photos by Paul Tulley.

The most sophisticated home-made traps involve two chambers to achieve these tasks. In addition, traps like these need to be easy to make and easy to maintain (top up with attractant, remove Asian hornets, let other insects go).

The JP33 design incorporates a top side entrance of 9 mm, followed by a funnel to the bait area, which is fitted with a 5.5-mm escape for non-target insects (see 'Resources'). It uses wooden tongue-depressors as guiding ramps for the entrance and the exit, and it doesn't use a separate roof. However, the 9-mm entrance allows European hornets to be caught. In Morbihan in 2017, along with 68,264 Asian hornets, 12,381 European hornets were caught (there is no information on the traps used or whether European hornets were released).

The National Bee Unit (NBU) has online instructions for making a monitoring bottle trap on the Asian hornet page of their website Bee Base (http://www.nationalbeeunit.com/index.cfm?sectionid=117) (the instructions download as a pdf; there is also a video).

This page and opposite: *Bob Tompkins' box trap made from a plastic storage box. Plastic funnels have had holes drilled in them and have been mounted on the short walls, and large areas of plastic grilles (queen excluders) have been mounted on the side walls to allow scent to leave the box and allow smaller insects to escape. The bait is in a tray beneath a mesh floor. There is a tight-fitting lid. Since other insects caught will be attracted to light, it would be good to darken the box except for the walls with escape grilles in them. Photos by Bob Tompkins*

The third type of trap is the box trap. Essentially, it is a much bigger version of the bottle trap, using a wooden or plastic box with funnel/cone entrances and sometimes escape holes for smaller insects. It's job is mass killing of Asian hornets. Attractant can be supplied soaked into a material (sponge, paper, cloth, wood shavings), or can be under a mesh floor, or the hive itself acts as the attractant (see later). The box can be used in an apiary to mop up Asian hornets that are preying on honey bees. The box trap is one of those responses to a problem that seems to have been thought up by several people and groups independently.

Raoul's cage

Raoul makes small portable boxes with clear perspex roofs. The Asian hornets enter by cones fashioned out of fine woven wire mesh — they can just squeeze in through the end hole and cannot get out. He baits the trap with sardines, old foundation and honey (not screened off), and catches around 300 hornets a day at the height of the season! He then kills them by hosing them with very hot water from his solar hot-water panels. (AAAFAe 2011)

The JABEPRODE

The JABEPRODE stands for Denis le **Ja**ffre's **be**e **pro**tection **de**vice. Maybe it's easier to say in French. Denis won the Lepine Competition grand prize in 2018 for his cones, which were originally made from expanded metal sheeting (the sort used in building work to hold render). Like Raoul, his cones emerge from a mesh wall, which allows good dispersion of bait odours. However, Denis' mesh walls have big enough holes that a lot of non-target insects can escape. He is in the process of getting a plastic version of his cones manufactured, which could then be mounted into a home-made box. In Denis' design, the bait is kept away from the Asian hornets by a mesh floor.

The ApiShield*

The ApiShield. The front is to the left, where bees enter a slot door and climb into the brood box through a short tunnel (opening seen at left). The bees are trained to use the front door before the side entrances for hornets are opened. The removable drawer is where hornets become trapped (you may just be able to make out red entrance cones inside the drawer. Photo by Vita Bee Health

*Disclosure: Vita Bee Health donated £200 towards producing this handbook.

One form of the box trap is the ApiShield, made by Vita Bee Health. Originally, it was patented as the Apiburg trap and was designed to trap wasps. Basically, it is a frame with an enclosed removable drawer (see photo). The frame has a horizontal slot entrance at the front of the hive, with a boxed-in tunnel for the bees to get up into the brood box. It has a varroa-mesh roof (which is therefore the floor of the colony space), an opening at the back that the drawer slides into, and four holes along each side. The drawer also has its own varroa-mesh roof, and has funnels (cones) in its sides, which line up with holes in the frame when the drawer is pushed home. Mesh-covered holes in the face of the drawer allow odours to circulate (including those of trapped hornets) and let in light, tricking the trapped insects into thinking that is the way out. The drawer traps the hornets and the whole drawer can be removed for inspection and hornet killing. Because the box relies on the odour and sound of the colony to attract the hornets, it doesn't need any other attractant. When it is first set up, the cones in the drawer are blocked off for a few days until the bees are familiar with the front entrance; then the cones are opened and hornets, bypassing the guards at the entrance, go through the side entrances and get trapped.

The ApiShield is an expensive piece of kit (£50), but it is low maintenance (after initial wood treatment), and Vita Bee Health recommend one fitted to the weakest of every four or five hives. A study on the device in France removed *Vespa velutina* trapped in it and measured their dry weight; interestingly, after 11th October, a heavier class of Asian hornets appeared, presumably new gynes attracted to the sweet insides of the hive (Papachristoforou et al. 2013). Unfortunately, the trap can also catch native hornets, but rarely (Sebastian Owen, pers. comm.). It works, and the advantage of this type of trap is that once it is installed, the drawer can be removed easily and the Asian hornets dealt with, without having to add or clean out attractant. It also gives you a varroa floor if you don't already have one.

The fourth type of trap is the sticky trap. Vita Bee Health make one called ApiProtect. It is only available where there are Asian hornets, so it is not yet available in the UK. Basically, it is a board coated with an extremely sticky glue. A special bait, or a live Asian hornet is used to attract Asian hornets onto the gluey surface. Unfortunately, anything landing or walking on the surface cannot be released and so is doomed. A beekeeper in France who tried it out found that, alongside Asian hornets, mice, lizards, European hornets and moths were caught, so stopped using it (Kevin Baughen, pers. comm.). If lizards get caught on it, small birds might also get trapped. I would rule this out.

Another type of glue trap is the SFERA: a yellow plastic ball, wrapped in cling film and suspended from a frame. It is coated in mouse glue and then a little cat food is used to attract the hornets. The hornets land and become stuck in the glue; when the ball is covered, the cling film can be peeled off and it can be set up again. Used by beekeepers to keep down numbers of Asian hornets in apiaries when under severe attack: very unselective.

Type of attractant

When we look at the natural history of *Vespa velutina*, we see that, theoretically, different foods should be more attractive at different times of the year (see graph p. 31). When the foundress wakes up from hibernation, she should seek sweet carbohydrate types of food: sap, nectar, and therefore sweet attractants should work best. Once her first batch of eggs hatch into larvae, she will need to find protein for them (insects, carrion), and will receive larval exudate from the larvae once they are old enough. She is also likely to take advantage of other sources of sweet liquid carbohydrates when she is out and about.

Once the first workers emerge from pupation, they will start foraging to get protein to feed the larvae in the nest. Some authors suggest swopping from a sweet attractant to a protein-based attractant (raw fish) from the beginning of August to mid October (France; Monceau & Thiéry, 2016). But, workers still find it hard to resist a sweet attractant (after all, they need sweet carbohydrates for energy); in Jersey I saw Asian hornets feeding on Suterra in a dish placed between two hives, even though active honey bee hawking was going on in the apiary at the same time. In the autumn, when the sexuals leave the nest, they first feed as much as possible on larval exudate, which is a perfect mix of carbohydrates and proteins for the adult insect. Autumn traps should concentrate on sweet bait once again. Protein in traps goes off and gets smelly. At some point it will no longer be attractive to hornets and will attract blowflies (e.g. bluebottles) instead. Fish and shrimps have been tried out with some success in France, Jersey and the UK.

Work is ongoing in designing an attractant that is based on Asian hornet pheromones. If this is successful, then a truly selective trap, which will only attract Asian hornets, can be manufactured: the holy grail of trapping. Cheng et al. (2017) found that Asian hornet venom acts as an alarm pheromone, which will attract other hornets, and beekeepers in France regularly use live trapped Asian hornets to attract others, even going so far as to get one to explore funnels destined for traps, so as to leave its odour on the plastic.

Wen et al. (2017) have discovered two compounds that act as sex pheromones and work is underway to turn these into stable pheromone trap attractants: commercial versions should arrive soon. Another possible time-point for pheromones might be when the nest is relocated: perhaps a pheromone is released by workers to call the queen?

Trapping at different times has different results

In France, spring trapping is done to capture foundresses, but there is no clear evidence that it works (see earlier). The timing is to catch them either before they start to build a nest, or before the first workers appear. Therefore, French beekeepers put traps out when the temperature has reached a consistent 13° C (AAAFAa 2015). In France, sources such as AAAFA say that spring traps should be taken down on 1st May to reduce capture of native hornets and other by-catch; others say the end of May (Blot & Debellescize 2017). The timing is based on the first workers emerging: this happens around 50 days (around 7 weeks) after the foundress starts her nest, so traps should be removed 6-7 weeks after the first foundresses are spotted. The queen will continue foraging alongside her worker offspring for 2 or 3 weeks before staying permanently inside the nest (apart from when the nest is relocated) (Martin 2017).

Summer trapping should only take place amongst hives when trying to reduce the impact of predation on honey bees, while reducing the by-catch. Shah and Shah (1991) in Kashmir, northern India, described using a fermented mixture of honey:water (1:1), half-filling a glass jar to make a killing trap for severe predation. In 10 days, with 12 jars set out in an apiary, they caught 11,483 Asian hornets (96 per jar per day), and the numbers of hornets hawking in front of hives went from 10-25 per hive to 0-3. They did not mention whether there was any by-catch. Some UK beekeepers already use a similar technique for killing wasps around apiaries.

Rome et al. (2011b), using a modified version of Shah and Shah's technique, found that apiary trapping close to hives using a honeycomb bait (made from melting a frame of comb in 1.5 litres water with 20 g honey and allowing the mixture to ferment for at least 3 days) could be moderately effective: where there was heavy predation (over 10 hornets in front of hives), 30-40% of the insects trapped were *Vespa velutina* (as opposed to non-target species), which is still pretty bad from a by-catch point of view. Flies and bees (presumably honey bees) were the main non-target insects caught, which is why it is important to place such traps as close to the hawking Asian hornets as possible.

Decante (2015) looked at the effects of apiary trapping of *Vespa velutina* on non-target insects and on colony loss. He tried out a series of different traps and different baits in apiaries under low and high predation pressure from Asian hornets. His conclusions were that the best trap overall was the dome-type (where the hornet enters fom underneath), baited earlier in the autumn with a fish bait, and later with Véto-Pharma bait (the Véto-Pharma trap with Véto-Pharma bait caught the most honey bees out of all modalities tested). However, he concluded that there was no protective effect of trapping on foraging, or on the development or survival of colonies.

Autumn/winter trapping is done to try to catch males and gynes (new queens) and is therefore done once the sexuals are emerging. We don't know exactly when this will be in the UK: in Jersey gynes started appearing in late September. Depending on how fast the winter comes, traps could be taken away perhaps in early December.

Although autumn trapping is preferred to spring trapping by ecologists (e.g. Monceau et al. 2013a), it generally appears to be the least popular form of trapping. The main argument for trapping at this time of year is that there are fewer insects on the wing, and therefore perhaps you could expect a lower by-catch. If traps based on mating pheromones become available, this will become the main trapping time.

But if I kill a foundress in the spring, that's one less colony and therefore I will stop huge numbers of insects (including honey bees) from being killed later in the year, right?

Trapping of foundresses in the spring ('spring queen trapping') does not seem to affect the number of nests found later in the year. There have been many studies (see earlier) that demonstrate this. Not every trapped foundress goes on to found a colony (remember the nests in Tours: out of 12 embryo nests, only three survived to make colonies).

The species that *Vespa velutina* catches in the summer are different from those caught in traps in the spring (see earlier, and traps catch insects all day and night, while Asian hornets only catch prey during the day): because spring trapping doesn't seem to affect numbers of colonies, the non-target species trapped in spring traps merely adds to the insects losses caused by *Vespa velutina* predation later in the year (Rojas-Nossa 2018).

I think I've found an Asian hornet in my bait station/monitoring trap —what should I do?

Don't lose it!

If it's in a trap, then put the whole trap into a plastic bag and put it in the freezer. After a couple of hours, the hornet will be very torpid, if not dead. You can then put the hornet on a surface and take some photos.

If you have an identification guide to hand (this book!), rule out it being a native European hornet or one of the other insects it may be confused with. If you are still pretty sure it's an Asian hornet, then send your best photos to alertnonnative@ceh.ac.uk, or use the Asian Hornet Watch App on your smartphone. You will also need to send a note about where you found it and your own contact details.

Then put the dead or near-dead hornet in a small container back in the freezer and leave overnight (after which it will definitely be dead): you may need to show it to the National Bee Unit if they arrive on your doorstep, or they may want you to post it to them.

If you can, take some photos of it in the trap or feeding at the bait station straight away (they are not aggressive when feeding and a long way from the nest).

If you are using a bait station, try to catch the hornet in a small pot or queen-catcher, or use a net to catch it and transfer it to a pot (remember that they can give a nasty sting; be very careful when using a net) then freeze it as described.

If you have nothing to hand to catch it right away, but you do have a camera or phone, then take some photos of it feeding at the bait station before you find something to put it in; if it has gone by the time you get back, be patient, it is quite likely to return.

If you would like some help with identification, contact your local Asian Hornet Action Team, which is part of your local Beekeepers' Association. To find out who is nearest, there is a map on the British Beekeepers' Association website (www.bbka.org). Hover over Asian Hornet on the top menu bar, then choose 'Asian Hornet Action Team map' from the drop-down menu. Then you can expand the map and click on one of the map pins to get contact details of your nearest team.

Trapping conclusions

At going to press (spring 2019), Asian hornets have not established in the UK, so at the moment we only need to monitor. To monitor for the arrival of Asian hornets, use bait stations if possible.

Traps that use food baits are unselective — even when they try to 'strain out' Asian hornets by the use of different entrance and exit holes. If not monitored regularly to release other insects, traps are likely to kill 98-99 other insects for each Asian hornet trapped (and that's in an area where Asian hornets are fully established).

If Asian hornets establish in the UK, go ahead and trap any attacking your apiary in the summer and early autumn — in that case use traps with sweet or protein bait in amongst your hives. Make sure that the traps are removed as soon as the pressure drops off (for example after any nearby nests have been destroyed). Box traps, which allow non-target insects to escape, are the most environmentally friendly, but destroying the nests of the Asian hornet colonies that are attacking your hives is the best solution.

A pheromone trap that is specific to *Vespa velutina* seems very close now: let's hope that it arrives before the Asian hornets do.

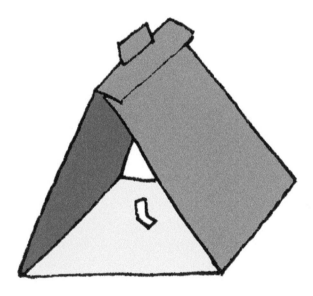

Drawing of a cardboard pheromone trap. The small white paper in the centre of the floor of the trap contains the pheromone; the yellow floor is a sticky trap.

Poisoning

Vespids (the social wasps, including hornets) have been successful in establishing all over the world: sometimes introduced by humans, on purpose or accidentally, sometimes spreading naturally across continents. The eusocial wasps (those that live in colonies, with only some capable of reproduction, like Asian hornets) seem to be particularly good at invading new areas; as we have seen, a single mated queen is all that is needed to establish in new territory, and their impact can be large and wide-ranging (Beggs et al. 2011).

So far, toxic baiting has been the most successful control strategy used against invasive vespids (Beggs et al. 2011), and now a poisonous bait has been designed with a stable shelf-life, it is being used on a much larger scale. Insecticide-laced bait has been successfully used to reduce populations of wasps all over the world (Patagonia in Argentina, Tasmania in Australia, Hawai'i in the USA, and in New Zealand). The insecticide used is fipronil, and it is mixed with minced meat at a concentration of 0.1%. Using the commercial version of this bait (Vespex®), traffic was reduced by 97% after around a month (Edwards et al. 2017). In their paper that looked at the transition from field trials to larger areas, they were able to bait plots from 217 ha to 2477 ha in New Zealand, against *Vespula* wasps. Nests were not removed after poisoning — indeed on these large scales, they didn't even know where they were. However, the New Zealand Department of Conservation managed the sites where the bait was tested and the New Zealand Environmental Protection Authority approved the use of Vespex® in New Zealand.

The problem in Europe is that there are plenty of scavengers (insects which will eat meat or meat juice) that are non-target insects, including wasps and the European hornet.

In France, people have resorted to mixing up a home-made poison, using fipronil, egg, flour and water, and painting it onto captured workers. Anecdotally, around 10 workers need to be treated in this way in order to kill the colony. The problem with this method is that the nest is not discovered and once it is undefended, birds, small mammals and other insects will move in to feed on dying adults, pupae and larvae, potentially causing a cascade of poisoning in the local ecosystem. In Spain, a similar method, which they call the 'Trojan method', involves spraying a worker with insecticide, removing its back legs to prevent it from grooming, and letting it go back to the nest.

Fipronil can take years to break down, especially if it is in the dark (half-life 4 months to 1 year): it is highly persistent in soils, tends to accumulate in soils and sediments, and has a high run-off and leaching potential to surface and groundwater (Pisa et al. 2014).

Fipronil is extremely toxic to bees. It is used as a soil treatment and seed dressing, but was banned from use on maize and sunflowers from the end of 2013 in Europe after a mass die-off of bees in France was blamed on fipronil-dressed seeds. Due to its nature and widespread use, Pisa et al. (2014) note that:

> *Despite large knowledge gaps and uncertainties, enough knowledge exists to conclude that existing levels of pollution with neonicotinoids and fipronil resulting from presently authorized uses frequently exceed the lowest observed adverse effect concentrations and are thus likely to have large-scale and wide ranging negative biological and ecological impacts on a wide range of non-target invertebrates in terrestrial, aquatic and benthic habitats.*

In Japan (Kishi & Goka 2017), where they are in a similar position to the UK in that they are expecting Asian hornets to establish on the mainland, they have been looking at chemicals that could be used against them. They investigated 22 possible compounds: of the six most effective chemicals, three are now prohibited in most countries! This left them with fipronil and diazinon: the former can harm important pollinators and other non-target invertebrates, while the latter is an organophosphate insecticide, suspected of being carcinogenic.

Looking for alternatives, they are investigating insect growth regulators (IGRs), which strongly affect the development of eggs, larvae and pupae, but do not kill adults immediately. They have had some success with etoxazole, which suppressed the pupation of Japanese native hornets in the laboratory and caused colonies to collapse in the field. They are continuing with field trials, but again, no-one knows the effects of these chemicals on non-target organisms.

....

Half-life: the time required for half of the compound to degrade.

1 half-life = 50% remaining
2 half-lives = 25% remaining
3 half-lives = 12% remaining
4 half-lives = 6% remaining
5 half-lives = 3% remaining

The amount of chemical remaining after a half-life will always depend on the amount of the chemical present initially.
Information from NPIC (National Pesticide Information Center, US Environmental Protection Agency)

Biological control

In their native Eastern China, Asian hornets have few natural predators: other hornets may attack a small or weak nest, and the European honey buzzard (*Pernis apivorus*) may take some (Martin 2017). By far the greatest killer of Asian hornets is humans — for food, protection of apiaries and protection of communities in Asia, or during eradication programmes in Europe.

Although their predators are few, we all have smaller things out to get us, and so attention has turned to European parasites that could help bring Asian hornets under control. When organisms are transplanted into a completely new environment, they may be freed from the parasites from their native range, but be susceptible to parasites in their new territory. They may be vulnerable to diseases and parasites carried by insects they come into contact with, especially honey bees, European hornets, wasps and other prey.

However, it may take time for native parasites to adapt to invasive species, and in the meantime, the pressure taken off alien species by leaving their parasites behind can mean that they can put more resources into growth and breeding: this may be one of the factors involved in the spectacular spread of Asian hornets through Europe (Darrouzet et al. 2014).

Conops vesicularis

In the summer of 2013, twelve primary Asian hornet nests were observed on buildings in and around Tours, central France (Darrouzet et al. 2014). Of these, only three developed normally, showing what a hazardous time foundation of colonies is for foundresses. Of the colonies that died out, seven had queens which could not be found, and two had dead queens in them. These queens were dissected and both were found to have been parasitised by an extremely specialist type of fly, *Conops vesicularis*,

Conops vesicularis. *Photo by By Tristram Brelstaff (Own work, CC BY-SA 3.0)*

native to Europe. So perhaps this fly will naturally help to control Asian hornets. Another possibility is to use these flies as biological control agents but, obviously, because they can attack non-target insects (they are known to attack bumblebees), such a use could be too risky to native insects to contemplate.

Other insect parasites

There are a few insects which parasitise hornets other than *Vespa velutina* that may be worth pursuing. *Xenos moutoni* (family Stylopidae) parasitises *Vespa analis* in Japan (Makino et al. 2011); when parasitised, the hornet is unable to reproduce.

Sphecophaga vesparum (family Ichneumonidae) has been found to parasitise *Vespa orientalis* in Israel (Turchi & Derijard 2018). The females enter the nest and lay their eggs on top of the host's larvae or pupae. The parasitic larva then attacks and feeds on the host's larva and then continues its development in a cocoon in the nest.

Volucella inanis is a hover-fly (family Syrphidae), and part of an amazing genus of hover-flies which mimic and parasitise different social hymenopterans. The larvae enter the cells of their host and eat their larvae.

Nematodes

In a 2015 paper, Villemant et al. discussed the possibility of mermithid nematodes being used as a biological control for Asian hornets. The only thing is that, in 10 years, only three nematodes have been found, despite scientists handling thousands of hornets (33,000 hornets from 77 nests). The species (perhaps *Pheromermis vesparum* — one which specialises in social wasps in Europe) matures in the abdomen of the host, and can be 8 cm long, which would render a gyne (queen) host sterile.

Another two species of nematodes, *Sphaerularia vespae* and *Sphaerularia bombi* have been found to parasitise gynes of *Vespa simillina*, resulting in the queens being unable to reproduce. In Japan, more than 60% of *Vespa simillina* gynes collected were infected with these nematodes (Kanzaki et al. 2007). Both of these nematodes are present in France, so there is potential for them to be used as biological controls, but they could also infect bumblebees.
Finally, *Steinernema feltiae* is a nematode which is already used as a biocontrol in France against caterpillars and ants and has been tested against wasps: there is some potential here, too (Turchi & Derijard 2018).

Entomopathogenic fungi

Entomopathogenic fungi are those that specialise in attacking insects: the fungal spores land on the insect and, if conditions are right (especially temperature and humidity), they can germinate and their hyphae can travel right through the insect, killing it in the process.

Poidatz et al. (2018b) have looked at using these fungi for Asian hornet control. They concluded that perhaps the best way to use the fungus would be to spray a suspension of spores into a nest, so replacing the broad-spectrum insecticides that are used for nest destruction at the moment. Although whether such a method could infect non-target insects needs more investigation, the authors think that this risk is low. In a more recent paper (Poidatz et al. 2019), they also discuss the possibility of infecting Asian hornets as they access bait; the infected insects then return to the nest and infect the colony.

Carnivorous plants

Christian Besson, gardener and botanist in the Nantes Botanical Gardens, France, first noticed many *Vespa velutina* were getting trapped by *Sarracenia* — a type of North American pitcher plant. Meurgey & Perrocheau (2015), from the Natural History Museum and Botanical Gardens, respectively, in Nantes, then cut open 203 pitchers to find out how many were being caught. Another study (Wycke et al. 2018) found that *Vespa velutina* were only 4.3% and 0.7% of all insects caught in pitchers in 2015 and 2016, respectively, although there were several Asian hornet nests nearby. The vast majority of insects caught were flies.

Sarracenia. *Photo by Noah Elhardt (CC BY-SA 3.0)*

So, not selective enough, and the plants themselves are tricky to cultivate.

Chickens

Job Opportunity — for self-confident chickens with good eye-beak co-ordination, there are several permanent positions available in an apiary near you! Some beekeepers in France have noticed that the free-range chickens which forage around their hives actually pluck Asian hornets out of the air when they are hawking in front of hives. I don't know how they avoid the venom from the hornets, or how many they can deal with in a day. But since the combination of orchard, apiary and chickens is quite common, it's worth a try. (video: Inform'Action 2015).

By the way, don't expect chickens to deal with a nest in an enclosed space — if the hornets build a nest in the chicken coop, a chicken caught in there with them can easily be stung to death (Kevin Baughen, pers. comm.).

Deformed wing virus

Forzan et al. (2017) found European hornet queens to be infected with deformed wing virus (DWV), a common disease of honey bees transmitted by the varroa mite (*Varroa destructor*). DWV has also been found on flower pollen and in pollen loads taken back to the honey bee hive. It may be that *Vespa crabro* picked up the virus by being fed an infected honey bee when it was a larva, but other routes cannot be ruled out.

Although this opens the possibility that Asian hornets may be able to pick up DWV from infected honey bees, they will then likely just mirror the level of infection in local honey bees, and beekeepers want to reduce levels of DWV, not increase them.

Mites

Very few mites are found in vespids (Turchi & Derijard 2018), but recently researchers in New Zealand have isolated *Pnemolaelaps niutirani* in their search for biological controls against wasps: perhaps a mite could be used against *Vespa velutina*.

. . . .

Robinet et al. (2016) suggest using combined strategies of mechanical removal and destruction of nests alongside biological control and pheromone traps to slow down the spread of *Vespa velutina* and reduce its population density at a large scale. They also point to the inherent riskiness of biological control, which can go horribly wrong due to unforeseen consequences.

Genetics

As we have seen, the ability of a single mated queen to start a colony, and therefore an invasion, has been highly beneficial for wasps and hornets. But such an event produces an extreme genetic 'bottleneck': a large amount of genetic diversity is lost.

Arca et al. (2015) looked at the French and Korean populations of *Vespa velutina* and found that they had much less genetic diversity than the native populations in China, indicating a severe genetic bottleneck caused by single founder events for these new invasions. Despite such a narrowing of genetic diversity, Asian hornets have conducted a spectacularly fast invasion of Europe, due to a favourable climate, easy food supplies in the form of the local honey bees (*Apis mellifera*), possibly a lack of parasites they would normally be exposed to, and lack of competition from local hornets.

So far, there isn't much evidence for detrimental effects of the bottleneck. One possible indicator is the production of diploid males: eggs that should have developed into females develop into males instead, due to a lack of diversity of genes for sex allocation. When a lot of diploid males are produced very early in the season, the colony may fail due to lack of help in feeding larvae and nest building.

Selection of honey bee strains

In Europe, the docility of honey bees is a top trait that is selected for, alongside colony health and productivity. Maybe we will have to rethink this, at least in terms of aggression towards Asian hornets. As we have seen, the honey bee prevalent in *Vespa velutina*'s home territory, *Apis cerana* displays defensive behaviour which helps the colony survive, including heat-balling, shimmering and different evasive flight patterns around the hive. Perhaps in the future we will have to consider selection of colonies which are able to cope with the pressure of Asian hornet predation (Monceau et al. 2014a).

DNA technology

A possible high-tech solution to *Vespa velutina* control would be to interfere with its gene expression, for example stopping it from producing some vital component used for its development. First, suitable genes to interfere with would need to be found and understood in the hornet, and then a delivery system would need to be developed. Because the larvae are fed meat, the production of a genetically modified fish, for example, which produced specific RNA interference molecules that would interfere with larval development, could be developed as a biologically toxic bait. Such an approach could be extremely specific (Turchi & Derijard 2018).

The development of a recessive infertility gene might be another avenue (see Hammond et al. 2016).

Turchi & Derijard (2018) note that these types of genetic modifications can have far-reaching consequences if they reach unintended populations. For example, if *Vespa velutina* carrying the modified infertility gene were to get into the native population back in China, it could have massive knock-on effects in the ecology there, where several different species of hornets at present co-exist.

Asian hornet seeking nectar from a Camellia sasanqua *Navajo, on Jersey. Photo by Peter Kennedy*

Hive defence

If Asian hornets become established in this country and start attacking honey bee colonies, the best way to defend your apiary would be to find and have destroyed any local nests (within 500 m, or even a kilometre radius if you can manage that). If that is not possible, then you will have to deal with the hornets as they arrive at the apiary.

We have mentioned trapping of foragers, especially with box traps, to reduce the numbers of hornets hawking. Traps in amongst hives are less likely to catch non-target insects than traps set to lure Asian hornets away from hives. We have also mentioned the possibility of chickens snacking on hornets. Do bear in mind, though, that one study (Decante 2015) found no increased survival of colonies in apiaries provided with Asian hornet traps compared to those without traps.

Reduction of entrances

A simple action to at least stop Asian hornets getting inside your hive is to use a guard (like a mouse guard) with holes no bigger than 6 mm. Kevin Baughen says that beekeepers in his part of France (between Limoges and Poitiers) use plastic anti-hornet entrance reducers in the autumn when colony numbers are dwindling, to prevent hornets from entering weaker colonies. They then stay on as mouse guards for the winter. Drones do struggle to get through the gaps (which are 5 mm high), so they are removed before drone build-up in the spring.

Disrupting hawking

Hawking (the Asian hornet behaviour where they hover in front of hives and attack flying honey bees) can be made difficult by allowing grass to grow long in front of the hive entrance, but as well as providing interference for hovering Asian hornets, if it forces bees to land away from the hive and carry on by foot, they can also become easy targets. By screening off the areas under hives, if they are up on stands, Asian hornets can be stopped from using these areas to wait for returning foragers. In addition, some kind of mesh, net or cage can be used to disrupt hawking at the hive front.

The holes in the mesh need to be big enough for bees to fly straight through, but small enough that hornets will hesitate because of the size of the gap and the fact that they would be entering the colony's territory. Don't be tempted to use a net or cage with smaller holes that the hornet can't get through, as that means the bees have to land on the mesh and then crawl through, making them an easy target.

The most popular netting arrangement is the 'museliere' or muzzle, a wire mesh contraption that covers the landing board of the hive. There are many variations on this because they are home-made. The typical mesh used is around 25 mm (1 inch): apparently chicken wire works just as well. Whether the muzzle has a solid floor and/or sides is up for experimentation: will more exits for the bees help?
Some beekeepers simply curl a piece of wire mesh from below the

This page and opposite: home-made 'muzzles'. These have solid sides and floors.
Photos by Bob Tompkins

landing board up to the roof, and tuck it under the roof; and I have seen whole hives under nets — but that would be quite awkward to deal with when inspecting hives.

Finally, another approach is the 'vertical muzzle', which in fact is a tunnel or 'chimney', 4 cm wide, which forces bees and hornets to fly down vertically to the hive entrance (Turchi & Derijard 2018), putting hornets into a vulnerable position where they could be attacked by guards.

Thwack and zap

Quite a few French beekeepers use swotting with a plastic raquet, followed by a judicious stamp to destroy hawking hornets, but if there are a few big nests in the area, you will hardly make an impact unless you are able to kill hundreds in a day, for days on end. The electric raquets designed to despatch mosquitos also look satisfying, but the hornet is only stunned and will again need to be stamped on to kill it.

The electric harp ('harpe electrique') is another way to kill Asian hornets in the apiary. It is a frame that contains several fine wires that are electrified. It is placed between the hives, behind the hives or in the flight path of the hornets. It is powered by a car battery charged by a solar panel with a specific controller. It straddles a tray filled with water: the hornet flies into the harp, becomes stunned by an electric shock, falls into the water and drowns. One YouTube video shows a catch after a week of 1,170 Asian hornets, no European hornets and 14 bees. In France the harp has won three awards. Electric harps can be bought or made at home: plans are available free on the internet (see 'Resources').

The electric harp.
The website is best found through 'apiprotect' on Facebook.

Feeding

Unfortunately, if colonies are suffering from foraging paralysis, they will need feeding. Again and again, reports from France emphasise the amount of extra feeding required to keep colonies alive.

How much is this going to cost us?

In a paper that came out in 2018 (Ferreira-Golpe et al.) beekeepers and bee farmers in the province of Coruña, Spain, were asked about what they were doing to counteract Asian hornets and how much these measures cost. The Spanish beekeepers used trapping most, followed by reducing the hive entrance size, nest destruction and the 'Trojan' method (poisoned insect goes back to nest). Only six farms used moving the hives (transhumance), three used electric harps and one used muzzles. There was also the cost of extra feeding to help colonies survive. All in all, they used 20% of the value of their production in combating Asian hornets.

Glossary

abdomen
The bulbous back half of a hornet (see anatomy drawing)

AHAT
Asian Hornet Action Team

APHA
Animal and Plant Health Agency, an executive agency of Defra

apiary
A collection of hives managed by a beekeeper

BBKA
British Beekeepers' Association

benthic
Environment on and in the sediment at the bottom of the sea, lakes, rivers etc

brood
Young life stages of honey bees and hornets (larvae, pupae)

callow
A freshly emerged (eclosed) adult (see teneral)

capture-mark-recapture
A technique used by biologists to estimate the number of organisms in an area

castes
The distinct forms that occur in a colony of a social insect: queen, worker, male

CEH
Centre for Ecology and Hydrology

cell
Hexagonal vessel in which a single larva/pupa is raised

colony
A collection of social insects with usually a single queen which act together in an organised way to raise offspring to produce the next generation

cocoon
In Asian hornets, the oldest stage of larva spins a cocoon around itself, which is proud of the paper cell, in which to pupate

comb
Short for honeycomb. The plates of hexagonal cells in which young are raised: horizontal and made of paper in wasps; vertical and made of wax in honey bees

cuticle
The skin of an insect — it may be thin and flexible, or thick where the exoskeleton needs to be strong.

Defra
Department for Environment, Food & Rural Affairs, a government department

diploid
Having two sets of each chromosome (twice that of haploid)

eclosion
Usually used by entomologists to describe the process of emerging from the pupa as an adult insect. Can also refer to a larva hatching from an egg

embryo nest
The initial Asian hornet nest built solely by the foundress queen

endocrine
To do with the internal secretion of hormones

entomofauna
The insects found in a particular place

entomologist
Someone who studies insects

ESA
Epidémiosurveillance santé animale: French public biosecurity organisation

exoskeleton
The system of strong plates linked by flexible membranes which makes up the outer part and support structure of an adult insect, like a suit of armour

exudate
Something which exudes from something else: in this case some kind of liquid oozing out of a tree

family
A rank or group used by taxonomists (those that work out the relatedness of organisms). It is above genus, which is above species (family e.g. Vespidae, genus e.g. *Vespa*, species e.g. *velutina*)

FDGDON
La Fédération Départementale des Groupements de Défense contre les Organismes Nuisibles

Fera Science Ltd.
Formerly the Food and Environment Research Agency; now a joint venture between Capita and Defra

field, the
Experiments done in natural(ish) environments as opposed to in the laboratory

flight-mill

Device consisting of a small stand with a freely-rotating wire arm to which the insect is tethered. It is used to measure speed of flight and lengths of time the insect is willing to fly

foundress
Asian hornet mated queen who has survived the winter and is now starting to establish a colony

FREDON
Fédération Régionale de Défense contre les Organismes Nuisibles

gaster
Another name for abdomen, used especially in the Hymenoptera

genus (pl. genera)
A group of closely related species, which all share the same first name, e.g. *Vespa* (it is always capitalized and should also be in italics)

gyne
A female sexual: once mated she will become a queen

haploid
Having a single set of unpaired chromosomes

Hymenoptera
An order of insects which includes bees, wasps, ants, sawflies etc.

IAS
Invasive alien species — category recognised by the European Union

INPN
Inventaire National du Patrimoine Naturel. Catalogue of natural biodiversity and geodiversity in France, managed by MNHN

INRA
L'Institut national de la recherche agronomique. French National Institute for Agricultural Research

instar
Stage of an insect between one moult and the next

larva, pl. larvae
In insects, it is the early immature stage, before pupation; in some insects they are known as maggots or grubs

mandibles
Pair of hard mouthparts used for biting, cutting or holding food. Also used in nest building in hornets

meconium (pl. meconia)
Waste from the gut expelled by hornet larva before it pupates

MNHN
Museum National d'Histoire Naturelle: French National Museum of Natural History

moult
Shedding of larval skin once it has become too tight; a new one lies beneath with more room for expansion

n
In academic biology articles, 'n' stands for number, so if n=9, then the data is from 9 individuals

NBU
National Bee Unit: part of APHA

nest
In hornets, the paper shelter built by the workers to house adults and young

NNSS
GB Non-native Species Secretariat

NPIC
National Pesticide Information Center, US Environmental Protection Agency

olfaction
The sense of smell or the act of smelling

petiole
The waist of the hornet, between the thorax and abdomen (see anatomical drawing opposite). Also the stalk which connects the combs to the supporting structure above them (see structures of primary and secondary nests)

pheromone
A secreted or excreted chemical factor that triggers a social response in members of the same species

primary nest
The first nest in hornets. They may relocate to a secondary nest once the colony is big enough

propolis
A red-brown, sticky, resinous material made by honey bees from material gathered from tree buds. It is used by the bees to fill gaps in hives among many other uses

pupa (pl. pupae)
Stage in the hornet between larva and adult, during which a complete transformation takes place from the larval form to the adult form. The change takes place hidden in the cocoon

queen
The sole reproductive female in a colony of hornets or honey bees. Also referred to as 'gyne' before mating, as 'foundress' when starting a new colony in the spring and 'mother queen' in a mature colony, to distinguish her from the young gynes which may also be present

RFID
Radio frequency identification, a

technology which uses tiny computer chips to track individual insects

secondary nest
If a colony of hornets relocates, the nest which is built is called the secondary nest

sexuals
Males (called 'drones' in honey bee terminology, but not in hornet terminology) and females which will mate with peers from other colonies. The female sexuals are called gynes in hornets and are destined to become queens

species
A type of organism. In *Vespa velutina nigrithorax*, '*Vespa*' is the genus name, '*velutina*' is the species name, and '*nigrithorax*' is the subspecies name. The species and subspecies names are never capitalized, but the whole name is usually in italics

superorganism
A colony of insects which function in some ways like a single organism (for example, keeping a stable temperature in the nest, the sexuals are analogous to sperm and eggs, etc)

tegula (pl. tegulae)
One of a pair of small hooded covers over the join between the wing and the thorax

teneral
Recently emerged adult (see callow)

thorax
Part of the adult insect between the head and the abdomen. It houses powerful flight muscles which work the wings. The legs are also attached to the thorax

trophallaxis
The transfer of regurgitated liquid food between adults or adults and larvae

UAV
Unmanned aerial vehicle, commonly called a 'drone'

worker
A usually non-reproducing female in a colony of hymenopteran insects. She will perform a range of different jobs (sometimes according to age) such as forager, guard, builder, cleaner

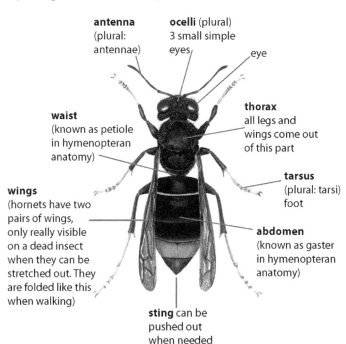

antenna (plural: antennae)

ocelli (plural) 3 small simple eyes

eye

waist (known as petiole in hymenopteran anatomy)

thorax all legs and wings come out of this part

wings (hornets have two pairs of wings, only really visible on a dead insect when they can be stretched out. They are folded like this when walking)

tarsus (plural: tarsi) foot

abdomen (known as gaster in hymenopteran anatomy)

sting can be pushed out when needed

Resources

13 Bees Website of Kevin and Amanda Baughen - Bee friendly holidays in France
www.13bees.co.uk

AAAFA Association Action Anti Frelon Asiatique
www.anti-frelon-asiatique.com

AHATs Asian Hornet Action Teams
www.ahat.org.uk/

BBKA British Beekeepers' Association
www.bbka.org

BBwear
www.bbwear.co.uk/

BWARS Bees, Wasps & Ants Recording Society
www.bwars.com

DIY plans for electrified harp or muzzle (in French)
http://mielleriedesgraves.over-blog.com/2015/09/plan-d-une-harpe-ou-museliere-electrique-anti-frelon.html

Electric harp - 'La Harpe Electrique' (In French)
www.apiprotection.fr/

FDGDON du Morbihan
La Fédération Départementale des Groupements de Défense contre les Organismes Nuisibles: http://www.fredon-bretagne.com/fdgdon-morbihan/

FREDON (Bretagne)
Fédération Régionale de Défense contre les Organismes Nuisibles
www.fredon-bretagne.com

French suppliers of beekeeping equipment (in French)
(Asian hornet excluder to go over hive door = 'portièr d'entree anti frélons')
www.apiculture.net

Jersey Asian Hornet Group facebook page
https://www.facebook.com/groups/293640477963172/
see also 'Jersey Asian hornet diary' on YouTube - superb videos from John de Carteret

Le blog de JP33 Blog and trap making (in French)
www.anti-frelon-d-asie-jp33.over-blog.com/article-piege-tres-selectif-pour-la-grande-serie-67774423.html

MNHN Muséum National D'Histoire Naturelle
Up-to-date map of European spread of Asian hornets
http://frelonasiatique.mnhn.fr/home/

NBU National Bee Unit
http://www.nationalbeeunit.com/
Trap instructions: www.nationalbeeunit.com/downloadDocument.cfm?id=1056

NNSS GB non-native secretariat
http://www.nonnativespecies.org/home/index.cfm

Northern Bee Books Specialist booksellers on all to do with bees
www.northernbeebooks.co.uk/

Paintball Asian hornet insecticide gun
www.frelons.com

UNAF Union National De L'Apiculture Française
www.unaf-apiculture.info/

uni POSCA marker pens
their website: www.posca.com, wide range from www.cultpens.com

Vita Bee Health - ApiShield box trap
www.vita-europe.com/beehealth/

Xesús Feás Spanish (Galicia) website with interesting research (in English)
www.vespavelutina.co.uk/

AHAT kit list

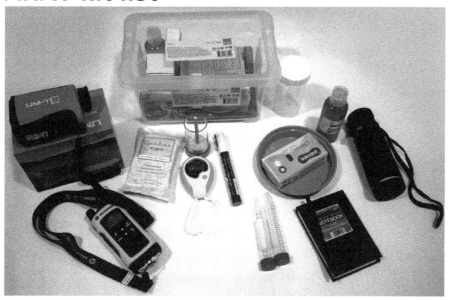

Several medium- to small-sized **plastic plant saucers**
Paper **kitchen roll** or similar
Small labelled **plastic bottles filled with Suterra**
Stopwatches
Compasses
uni POSCA marker pens (PC-3M), several colours (light colours are best)
Small **notebooks** and **pens**
Plastic tubes with screw lids (around 30 mm diameter, 45 ml capacity)
Yellow builders' buckets
Walkie-talkies (make sure different AHAT teams use compatible models so that if teams work together, they can listen AND talk to one another). If you buy ones with integral rechargable batteries and these are not kept charged over the winter, you will ruin the batteries, and possibly the walkie-talkie. If you use removable rechargable batteries which are removed over the winter you can avoid this.
Rechargeable batteries (plenty) plus charger (if that's what your walkie-talkies use).
Binoculars (usually a personal item, but may need to buy)
Rangefinder (optional) This is an expensive piece of kit (around £60). It is useful to find out how far away things are. You can point it at a clump of trees (it will give you a reading of the distance) to find out whether the distance to them ties in with the return times you are getting from market hornets. Instead, you can do an exercise with the team to pace out distances to objects to get a feel for how far away things are.
Sturdy container to keep everything in.

References

AAAFAa (2015) Association Action Anti Frelon Asiatique. Cycle de vie d'une colonie et piégeage de printemps, under Biologie. https://anti-frelon-asiatique.com/biologie/cycle-de-vie-dune-colonie-et-piegeage-de-printemps/ (Accessed 8 January 2019)

AAAFAb (2016) Association Action Anti Frelon Asiatique. Capture Reine dans Camélia - Pince à passoires. https://vimeo.com/156297742 (Accessed 19 April 2019)

AAAFAc (2016) Association Action Anti Frelon Asiatique. The jar-and-card technique (video). https://anti-frelon-asiatique.com/piegeage/detruire-un-jeune-nid-sans-insecticides-et-pour-100-mag-m6/ (Accessed 6 March 2019)

AAAFAd (2015) Association Action Anti Frelon Asiatique. Sieves on a litter-picker technique (video). https://anti-frelon-asiatique.com/piegeage/recuperation-100-naturelle-dun-nid-primaire/ (Accessed 6 March 2019)

AAAFAe (2011) Association Action Anti Frelon Asiatique. La cage de Raoul (video). https://anti-frelon-asiatique.com/piegeage-2/la-cage-de-raoul-piege-a-frelons-asiatiques/ (Accessed 9 February 2019)

Arca M, Papachristoforou A, Mougel F, Rortais A, Monceau K, Bonnard O, Tardy P, Thiéry D, Silvain J-F, Arnold G (2014) Defensive behaviour of *Apis mellifera* against *Vespa velutina* in France: testing whether European honeybees can develop an effective collective defence against a new predator. Behavioural Processes 106, 122-129

Arca M, Mougel F, Guillemaud T, Dupas S, Rome Q, Perrard A, Muller F, Fossund A, Capdevielledula C, Torres-Leguizamon M, Chen XX, Tn JL, Jung C, Villemant C, Arnold G, Silvain JF, (2015) Reconstructing the invasion and the demographic history of the yellow-legged hornet, *Vespa velutina*, in Europe. Biol. Invasions, 17: 2357-2371

Barbet-Massin M, Rome Q, Muller F, Perrard A, Villemant C, Jiguet F (2013) Climate change increases the risk of invasion by the yellow-legged hornet. Biol Conserv 157:4-10

Beggs JR, Brockerhoff EG, Corley JC, Kenis M, Masciocchi M, Muller F, Rome Q, Villemant C (2011) Ecological effects and management of invasive Vespidae. Biocontrol 56: 505-526

Bendiocarb general fact sheet (2002) National Pesticide Information Center, USA. http://npic.orst.edu/factsheets/bendiogen.pdf. (Accessed 27 January 2019)

Blanchard P, Schurr F, Celle O, Cougoule N, Drajnudel P, Thiéry R, Faucon J-P, Ribiére M (2008) First detection of Israeli acute paralysis virus (IAPV) in France, a dicistrovirus affecting honeybees (*Apis mellifera*). J. Invertebr. Pathol. 99: 348-350

Blot J (2007) Le frelon asiatique (*Vespa velutina*). Le piégeage des fondatrices. Fiche technique apicole. Bull. Tech. Apic. 34 (4) 201-204 (in French)

Budge GE, Hodgetts J, Jones EP, Ostojá-Starzewski JC, Hall J, Tomkies V, Semmence N, Brown M, Wakefield M, Stainton K (2017) The invasion, provenance and diversity of *Vespa velutina* Lepeletier (Hymenoptera: Vespidae) in Great Britain. PLoS ONE 12(9)

BWARS Bees, Wasps & Ants Recording Society. www.bwars.com (Accessed 17 February 2019)

Carisio, L, Manino A, Lioy S, Sottosanti G & Porporato M (2018). Is *Vespa velutina* a threat to wild bee communities and ecosystem pollination services? PAGEPress publication (PV) Italy. Proceedings of XI European Congress of Entomology, (pp. 129). Naples, Italy

Carpenter JM, Kojima J (1997) Checklist of the species in the subfamily Vespinae (Insecta: Hymenoptera: Vespidae) Natural History Bulletin of Ibaraki University 1: 51-92

Chauzat M, Martin S (2009). A foreigner in France, biological information on the Asian hornet *Vespa velutina*, a recently introduced species. Biologist 56: 86-91

Cheng Y, Wen P, Dong S, Tan K, Nieh JC (2017) Poison and alarm: the Asian hornet uses sting venom volatiles as an alarm pheromone. Journal of Experimental Biology 220: 645-651

Chibrac J (2013) Un drone lutte contre les frelons asiatiques. Video. https://www.youtube.com/watch?v=KTkQ2fJfXyU

Chinery M (1986) Collins guide to the insects of Britain and Western Europe. Collins, London

Choi MB, Martin SJ, Lee JW (2012) Distribution, spread, and impact of the invasive hornet *Vespa velutina* in South Korea, Journal of Asia-Pacific Entomology, 15 (3): 473-477

Cini A, Cappa F, Petrocelli I, Pepiciello I, Bortolotti L, Cervo R (2018) Competition between the native and the introduced hornets *Vespa crabro* and *Vespa velutina*: a comparison of potentially relevant life-history traits. Ecological Entomology 43 (3) 351-362

Couto A, Monceau K, Bonnard O, Thiéry D, Sandoz J-C (2014) Olfactory attraction of the hornet *Vespa velutina* to honeybee colony odors and pheromones. PLoS ONE 9 (12)

Darrouzet E, Gévar J, Dupont S (2014) A scientific note about a parasitoid that can parasitize the yellow-legged hornet, *Vespa velutina nigrithorax*, in Europe. Apidologie 46: 130-132

Darrouzet E, Gévar J, Guignard Q, Aron S (2015) Production of early diploid males by European colonies of the invasive hornet *Vespa velutina nigrithorax*. PLoS ONE 10(9): e0136680

Dauphin P, Thomas H (2009) Quelques données sur le contenu des "piéges à frelons asiatiques" posés à Bordeaux (Gironde) en 2009. Bull. Soc. Linn. Bordeaux, Tome 144 (N.S.) no. 37 (3) 287-297 (in French)

Decante, D (2015) Lutte contre le frelon asiatique *Vespa velutina*: Évaluation comparative des modalités de piégeage de protection du rucher. http://itsap.asso.fr/wp-content/uploads/2016/03/cr_ evaluation) piegeage_vvelutina)2014.pdf (Arrives in downloads as a pdf. Accessed 18 February 2019) (in French)

Dong D, Wang Y (1989) A preliminary study on the biology of wasps *Vespa velutina auraria* Smith and *Vespa tropica ducalis* Smith (Hymenoptera: Vespidae). Zoological Research 10(2): 155-162

Edwards R (1980) Social wasps: their biology and control, The Rentokil Library, East Grinstead

Edwards E, Toft R, Joice N, Westbrooke I (2017) The efficacy of Vespex® wasp bait to control *Vespula* species (Hymenoptera: Vespidae) in New Zealand. International Journal of Pest Management 63: 266-272

FDGDON-Morbihan(2018)http://www.fredon-bretagne.com/fdgdon-morbihan/frelon-asiatique/synthese-2018/ (Accessed 22 April 2019)

Ferreira-Golpe MA, García Arias AI, Pérez-Fra M (2018) Costes de la lucha contra la especie invasora *Vespa velutina* soportados por los apicultores en la provincia de a Coruña. Conference paper, CIER XII, Segovia 4th, 5th and 6th July 2018. (in Spanish)

Forzan M, Sagona S, Mazzel M, Felicioli A (2017) Detection of deformed wing virus in *Vespa crabro*. Bulletin of Insectology 70 (2): 261-265

Franklin DN, Brown MA. Datta S., Cuthbertson AGS, Budge GE, Keeling MJ (2017) Invasion dynamics of Asian hornet, *Vespa velutina* (Hymenoptera: Vespidae): a case study of a commune in south-west France. Appl Entomol Zool 52: 221

Gévar J, Bagnères A-G, Christidès J-P, Darrouzet E (2017) Chemical heterogeneity in inbred European population of the invasive hornet *Vespa velutina nigrithorax*. J Chem Ecol 42: 763-777

Gourbière S, Menu F (2009). Adaptive dynamics of dormancy duration variability: evolutionary trade-off and priority effect lead to suboptimal adaptation. Evolution: International Journal of Organic Evolution. 63. 1879-92.

Hammond A, Galizi R, Kyrou K, Simoni A, Siniscalchi C, Katsanos D...Nolan, T (2016) A CRISPR-Cas9 gene drive system targeting female reproduction in the malaria mosquito vector *Anopheles gambiae*. nature Biotechnology 34(1) 78-83. https://doi.org/10.1038/nbt.3439

de Haro L, Labadie M, Chanseau P, Cabot C, Blanc-Brisset I, Penouil F, National Coordination Committee for Toxicovigilance (2010) Medical consequences of the Asian black hornet (*Vespa velutina*) invasion in Southwestern France. Toxicon 55: 650-652

Haxaire J, Bouguet J-P and Tamisier J-P (2006) *Vespa velutina* Lepeletier, 1836, une redoutable nouveauté pour la faune de France (Hym., Vespidae). Bulletin de la Société entomologique de France 111: 194 (in French)

Haxaire J , Villemant C (2010) Impact sur l'entomofaune des "piéges à frelon asiatique". Insectes no. 159 (4) (in French)

Inform'Action (2015) Chickens eating Asian hornets (video). https://www.youtube.com/watch?v=JfQF8HxZqXY

JP33a (2011) Le Blog de JP33. Ongoing blog supporting eradication of Asian hornets, especially through spring trapping. Has designed successively more selective traps. http://anti-frelon-d-asie-jp33.over-blog.com/article-piege-tres-selectif-pour-la-grande-serie-67774423.html (Accessed 12 January 2019: in French)

JP33b (2011) Trap used by Monceau et al. in 2012. http://anti-frelon-d-asie-jp33.over-blog.com/article-piege-a-femelles-fondatrices-tres-selectif-plan-photo-65214764.html

Kanzaki N, Kosaka H, Sayama K, Takahashi J, Makino S (2007) *Sphaerularia vespae* sp. nov. (Nematoda, Tylenchomorpha, Sphaerularioidea), an endoparasite of a common Japanese hornet, *Vespa simillima* Smith (Insecta, Hymenoptera, Vespidae). Zoological Science, 24(11), 1134-1142. https://doi.org/10.2108/zsj.24.1134

Keeling MJ, Franklin DN, Datta S, Brown MA, Budge GE (2017) Predicting the spread of the Asian hornet (*Vespa velutina*) following its incursion into Great Britain. Scientific Reports (7) 6240

Kennedy PJ, Ford SM, Poidatz J, Thiéry D, Osborne JL (2018) Searching for nests of the invasive Asian hornet (*Vespa velutina*) using radio-telemetry. Communications Biology 1 (88)

Kishi S, Goka K (2017) Review of the invasive yellow-legged hornet, *Vespa velutina nigrithorax* (Hymenoptera: Vespidae), in Japan and its possible chemical control. Appl Entomol Zool 52: 361-368

Lui Z, Chen S, Zhou Y, Xie C, Zhu B, Zhu H, Liu S, Wang W, Chen H, Ji Y (2015) Deciphering the venomic transcriptome of killer-wasp *Vespa velutina*. Scientific Reports 5: 9454 DOI: 10.1038/srep09454

Makino S, Kawashima M, Kosaka H (2011) First record of occurrence of *Xenos moutoni* (Strepsiptera: Stylopidae), an important parasite of hornets (Hymenoptera: Vespidae: *Vespa*), in Korea. Journal of Asia-Pacific Entomology 14(1) 137-139. https://doi.org/10.1016/j.aspen.2010.09.001

Marris G, Brown M, Cuthbertson AG (2011) GB non-native organism risk assessment for *Vespa velutina nigrithorax*. http//www.nonnativespecies.org. (Accessed 12 Jan 2019)

Martin SJ (1990) Nest thermoregulation in *Vespa simillima*, *V. tropica* and *V. analis*. Ecological Entomology 15: 301-310

Martin SJ (1992) Colony defence against ants in *Vespa*. Insectes Sociaux 39, 99-113

Martin SJ (1995) Hornets of Malaysia. Malayan Nature Journal 49: 71-82

Martin SJ (2017) The Asian hornet — threats, biology and expansion. IBRA and Northern Bee Books, Hebden Bridge, England

Matsuura M, Yamane S (1990) Biology of vespine wasps. Springer-Verlag, Berlin

Meurgey F, Perrocheau R (2015) Les Sarracénies piéges pour le Frelon à pattes jaunes. Insectes no. 177 (2) (in French)

Milanesio D, Saccani M, Maggiora R, Laurino D, Porporato M (2016) Design of an harmonic radar for the tracking of the Asian yellow-legged hornet. Ecology and Evolution 6 (7)

Milanesio D, Saccani M, Maggiora R, Laurino D, Porporato M (2017) Recent upgrades of the harmonic radar for the tracking of the Asian yellow legged hornet. Ecology and Evolution 7 (13)

Mollet T, De la Torre C (2006) *Vespa velutina* — the Asian hornet. Bulletin Technique Apicole 33(4), 203-208. Translation in Bee Craft September 2007, 11-14

Monceau K, Bonnard O, Thiéry D (2012) Chasing the queens of the alien predator of honeybees: a water drop in the invasiveness ocean. Open J Ecol 2:183-191

Monceau K, Maher N, Bonnard O, Thiéry D (2013a) Predation pressure dynamics study of the recently introduced honeybee killer *Vespa velutina*: learning from the enemy. Apidologie 44: 209-221

Monceau K, Bonnard O, Thiéry D (2013b) Relationship between the age of *Vespa velutina* workers and their defensive behaviour established from colonies maintained in the laboratory. Insect Soc 60: 437-444

Monceau K, Arca M, Leprêtre L, Mougel F, Bonnard O, Silvain J-F, Maher N, Arnold G, Thiéry D (2013c) Native prey and invasive predator patterns of foraging activity: the case of the yellow-legged hornet predation at European honeybee hives. PLoS One 8 (6)

Monceau K, Bonnard O, Thiéry D. (2014a) *Vespa velutina*: a new invasive predator of honeybees in Europe. J Pest Sci 87:1-16

Monceau K, Bonnard O, Moreau J, Thiéry D (2014b) Spatial distribution of *Vespa velutina* individuals hunting at domestic honeybee hives: heterogeneity at a local scale. Insect Science 21, 765-774

Monceau K, Maher N, Bonnard O, Thiéry D (2015) Evaluation of competition between a native and an invasive hornet species: Do seasonal phenologies overlap? Bulletin of Entomological Research 105, 462-469

Monceau K, Thiéry D (2016) *Vespa velutina*: current situation and perspectives. Atti Accademia Nazionale Italiana di Entomologia Anno LXIV 137-142

Monceau K, Thiéry D (2017) *Vespa velutina* nest distribution at a local scale: an eight-year survey of the invasive honeybee predator. Insect Science. 24: 663-674

Monceau K, Tourat A, Arca M, Bonnard O, Arnold G, Thiéry D (2017) Daily and seasonal extranidal behaviour variations in the yellow-legged hornet, *Vespa velutina* Lepeletier (Hymenoptera: Vespidae). J Insect Behaviour 30 (2) 220-230

Papachristoforou A, Arca M, Ifantidid MD (2013) Trapping Asian hornets (*Vespa velutina*) workers and queens in France, using the Apiburg® hive bottom board. In: Proceedings of the XXXXIII International Apicultural Congress, 29th Sept-4th October 2013, Kyiv, Ukraine

Pawlyszyn B (1992) Nest relocation in the British Hornet *Vespa crabro gribodoi* Bequaert (Hym., Vespidae). Entomologist's Monthly Magazine 128: 203-205

Pérez-de-Heredia I, Darrouzet E, Goldarazena A, Romón P, Iturrondobeitia J-C (2017) Differentiating between gynes and workers in the invasive hornet *Vespa velutina* (Hymenoptera, Vespidae) in Europe. Journal of Hymenoptera 60: 119-133

Perrard A, Haxaire J, Rortais A, Villemant C. (2009) Observations on the colony activity of the Asian hornet *Vespa velutina* Lepeletier 1836 (Hymenoptera: Vespidae: Vespinae) in France. Ann. soc. entomol. Fr. (n.s.) 45 (1) : 119-127

Pisa LW, Amaral-Rogers V, Belzunces LP, Bonmatin J-M, Goulson D, Kreutzweiser D, Krupke C, Liess M, McField M, Morrissey C, Noome DA, Settele J, Simon-Delso N, Stark J, van der Sluijs, van Dyck H, Wiemers M (2014) Effects of neonicotinoids and fipronil on non-target invertebrates. Environ Sci Pollut Res 22 (1): 68–102

Poidatz J (2017) PhD Thesis. De la biologie des reproducteurs au comportement d'approvisionnement du nid,vers des pistes de biocontrôle du frelon asiatique *Vespa velutina* en France. Ecologie, Environnement. Université de Bordeaux (in French)

Poidatz J, Bressac C, Bonnard O, Thiery D (2017) Delayed sexual maturity in males of *Vespa velutina*. Insect Science. 10.1111/1744-7917.12452.

Poidatz J, Monceau K, Bonnard O, Thiéry D (2018a) Activity rhythm and action range of workers of the invasive hornet predator of honeybees *Vespa velutina*, measured by radio frequency identification tags. Ecology and Evolution 1-11

Poidatz J, Plantey RL, Thiéry D (2018b) Indigenous strains of *Beauveria* and *Metharizium* as potential biological control agents against the invasive hornet *Vespa velutina*. Journal of Invertebrate Pathology 153: 180-185

Poidtaz J, Plantey RJL, Thiéry D (2019) *Beauveria bassiana* strain naturally parasitizing the bee predator *Vespa velutina* in France. Entomologia Generalis March 2019. In press.

Requier F, Rome Q, Chiron G, Decante D, Marion S, Menard M, Muller F, Villemant C, Henry M (2018) Predation of the invasive Asian hornet affects foraging activity and survival probability of honey bees in Western Europe. Journal of Pest Science. 1612-4766

Reynaud L, Guérin-Lassous I (2016) Design of a force-based controlled mobility on aerial vehicles for pest management. Ad Hoc Networks, Elsevier, 53, pp.41 - 52. <10.1016/j.adhoc.2016.09.005>. <hal-01427874>

Robinet C, Suppo C and Darrouzet E (2017) Rapid spread of the invasive yellow-legged hornet in France: the role of human-mediated dispersal and the effects of control measures. J Appl Ecol, 54: 205-215

Rojas-Nossa SV, Novoa N, Serrano A, Calviño-Cancela M (2018) Performance of baited traps used as control tools for the invasive hornet *Vespa velutina* and their impact on non-target insects. Apidologie 49: 872.

Rome Q, Muller F, Olivier G, Villemant C (2009) Bilan 2008 De l'invasion de *Vespa velutina* Lepeletier en France (Hymenoptera: Vespidae). Bulletin de la Société entomologique de France, 114 (3): 297-302 (in French with English summary)

Rome Q, Perrard A, Muller F, Villemant C (2011a) Monitoring and control modalities of a honeybee predator, the yellow-legged hornet *Vespa velutina nigrithorax* (Hymenoptera: Vespidae). Aliens 31: 7-15

Rome Q, Muller F, Théry T, Andrivot J, Haubois S, Rosenstiehl E, Villemant C (2011b) Impact sur l'entomofaune des pièges à bière ou à jus de cirier utilisés dans la lutte contre le frelon asiatique. In: Barbançon, J-M, L'Hostis, M (eds). Journée Scientifique Apicole JSA, Arles, 11 février 2011. ONIRIS-FNOSAD, Nantes pp. 18-20 (in French)

Rome Q, Sourdeau C, Muller F, Villemant C (2013) Le piégeage du frelon asiatique *Vespa velutina nigrithorax*. Intérêts et dangers. Conference paper. Journées Nationales GTV - Nantes 2013 (in French)

Rome Q, Muller FJ, Touret-Alby A, Darrouzet E, Perrard A, Villemant C (2015) Caste differentiation and seasonal changes in *Vespa velutina* (Hym.: Vespidae) colonies in its introduced range. Journal of Applied Entomology. 2015;139(10):771-82

Rome Q, Villemant C (2018) Le Frelon asiatique *Vespa velutina*. In: INPN-MNHN. http://frelonasiatique.mnhn.fr/home. (Accessed 15 March 2019)

Sánchez-Bayo F, Wyckhuys K A G (2019) Worldwide decline of the entomofauna: a review of its drivers. Biological Conservation 232: 8-27

Sauvard D, Imbault V, Darrouzet E (2018) Flight capacities of yellow-legged hornet (*Vespa velutina nigrithorax*, Hymenoptera: Vespidae) workers from an invasive population in Europe. PLoS ONE 13(6)

Shah F, Shah T (1991) *Vespa velutina*, a serious pest of honey bees in Kashmir. Bee World 72, 161-164

Shearwood J, Hung DMY, Palego C, Cross P (2017) Energy harvesting devices for honey bee health monitoring. IEEE MTT-S International Microwave Workshop Series on Advanced Materials and Processes for RF and THz Applications (IMWS-AMP) 1-3, doi:10.1109/IMWS-AMP.2017.8247435

Spradbery JP (1973) Wasps: an account of the biology and natural history of social and solitary wasps. University of Washington Press, Seattle

Tan K, Hu Z, Chen W, Wang Z, Wang Y, et al. (2013) Fearful foragers: honey bees tune colony and individual foraging to multi-predator presence and food quality. PLoS ONE 8(9): e75841. doi:10.1371/journal.pone.0075841

Tan K, Radloff SE, Li JJ, Hepburn HR, Yang MX, Zhang LJ, Neumann P (2014) Bee-hawking by the wasp, *Vespa velutina*, on the honeybees *Apis cerana* and *A. mellifera*. Naturwissenschaften 94: 469-472

Thiéry D, Bonnard, O, Maher N, Poidatz J, Monceau K (2014) Comportement de prédation du frelon asiatique à pattes jaunes (*Vespa velutina*) et protection des ruches par différentes stratégies de piégeage. Conférence Ravageurs et insectes émergents invasif, AFPP 2014, Montpellier France. (in French with English Abstract)

Troadec N (2015) Finistère. Un homme décède après deux piqures de frelon asiatique. https://www.ouest-france.fr/bretagne/finistere-il-decede-apres-deux-piqures-de-frelon-asiatique-3614000. (Accessed 26 January 2019). (in French)

Turchi L, Derijard B (2018) Options for the biological and physical control of *Vespa velutina nigrithorax* (Hym.: Vespidae) in Europe: a review. Journal of Applied Entomology 142: 553-562

Ueno T (2015) Flower-visiting by the invasive hornet *Vespa velutina nigrithorax* (Hymenoptera: Vespidae). International Journal of Chemical, Environmental & Biological Sciences (IJCEBS) Volume 3, Issue 6

UNAF (2017) Asian hornet: spring trapping or not? UNAF takes stock. https://www.unaf-apiculture.info/actualites/frelon-asiatique-piegeage-de-printemps-ou-non-l-unaf-fait-le-point.html#nb1 (in French: accessed 13 November 2018)

Van der Vecht J (1957) The Vespinae of the Indo-Malayan and Papuan areas (Hymenoptera, Vespidae). Rijksmuseum van Natuurlijke Historie, Leiden

Villemant, C, Haxaire, J, Streito, J-C (2006) Premier bilan de l'invasion de *Vespa velutina* Lepeletier en France (Hymenoptera, Vespidae). Bull. Soc. Entomol. Fr. 111, 535–538 (in French, summary in English)

Villemant C, Barbet-Massin M, Perrard A, Muller F, Gargominy O, Jiguet F and Rome Q (2011a) Predicting the invasion risk by the alien beehawking yellow-legged hornet *Vespa velutina nigrithorax* across Europe and other continents with niche models. Biological conservation doi:10.1016/j.biocon.2011.04.009

Villemant C, Muller F, Haubois S, Perrard A, Darrouzet E, Rome Q (2011b) Bilan des travaux (MNHN et IRBI) sur l'invasion en France de *Vespa velutina*, le frelon asiatique prédateur d'abeilles. In: BarbanCon J-M, L'Hostis M (eds). Journée Scientifique Apicole JSA, Arles, 11 février 2011. ONIRIS-FNOSAD, Nantes, 3-12 (in French)

Villemant C, Zuccon D, Rome Q, Muller F, Poinar G, Justine J-L (2015) Can parasites halt the invader? Mermithid nematodes parasitizing the yellow-legged Asian hornet in France. PeerJ 3:e947; DOI 10.7717/peerj.947

Wen P, Cheng Y, Dong S, Wang Z, Tan K, Nieh J (2017) The sex pheromone of a globally invasive honey bee predator, the Asian eusocial hornet, *Vespa velutina*. Scientific Reports volume 7, Article number: 12956

Wycke M-A, Perrocheau R, Darrouzet E (2018) *Sarracenia* carnivorous plants cannot serve as efficient biological control of the invasive hornet *Vespa velutina nigrithorax* in Europe. Rethinking Ecology 3: 41-50

Yanez O, Zheng H-Q, Hu F-L, Neuman P, Dietemann V (2012) A scientific note on Israeli acute paralysis virus infection of Eastern honeybee *Apis cerana* and vespine predator *Vespa velutina*. Apidologie 43: 587- 589

Acknowledgements

My experience in Jersey was the impetus to writing this book, so special thanks to the welcoming fellow beekeepers there: Bob Hogge, entomologist, first tracker and gracious host; John de Carteret, whose photos and videos are informing people worldwide; fellow hornet hunters Judy Collins, Monique Pierce, Bob & Gill Tompkins and Nigel Errington, who all helped me with this handbook.
Simon O'Sullivan, Judith Norman, Gerry Stuart, Sue Baxter and Lynne Ingram, all beekeepers who travelled out to Jersey last season, have also helped with the handbook; Sue and Judith especially with the Tracking section, Lynne with the Asian hornet's life cycle, Gerry with AHAT material and Simon with general support for the project while trying to get the word out about Asian hornets via local events. My local Okehampton Beekeeping branch has given me wonderful support (especially Sue, Simon and Marian Minton), and Tony Lindsell (Devon Beekeepers) and Ken Edwards (Quantock Beekeepers) also went out of their way to help me get the project off the ground.

While we were in Jersey, Dr Peter Kennedy and Dr Jess Knapp (University of Exeter) were there too, with their cool off-road entomology lab. They were using radio telemetry to track hornets to their nests — fascinating and very useful. Peter Kennedy has been very kind in commenting on various drafts of this book, and in discussing aspects of Asian hornet biology, ecology and behaviour. Dr Quentin Rome (MNHN) and Dr Denis Thiéry (INRA) have also been kind enough to answer my questions.

I would also like to thank Colin Lodge, who wrote about his experience of instigating Asian hornet action teams; Kevin Baughen and Richard Noel who both keep bees in different parts of France and could thus relate their experiences of dealing with Asian hornets on a day-to-day basis; Robert Moon, pest controller in France, and Steve Bright from BBwear, who also answered many of my questions; Rev. Dr A Wakeham-Dawson, editor of Entomologist's Monthly Magazine for helping locate an article for me; Vita Bee Health for a £200 donation towards producing this book (but who have not influenced the contents in any way); Jonathan How, layout mentor; and all those who have allowed me to use their photos (you are credited by your photos — thank you — this book would be very dull without pictures! Any unattributed photos or illustrations are the author's). Thanks too to Robin Wootton, Clive Betts and Jeanette Mitchell for proof-reading: any remaining errors are my responsibility.

Finally, thanks to my partner Paul who has become an Asian hornet widow these last few months, while I have spent every spare minute on this book: no crazy deadlines next time, darlin'!

Index

A

AAAFA 109, 119
Accepted by other colonies 26
Acknowledgements 149
Activities 18
Adult diet 27
Aerial map 71
Aethina tumida 61
AHATs (Asian Hornet Action Teams) 57, 59, 60, 65, 70
  AHAT kit list 139
  fourfold aims 67
Alarm pheromone 44, 118
Alderney 50
Andernos-les-Bains 13, 22, 42, 106
Antennae 9, 75
APHA 60, 65, 69
Apiary predation 31
Apiary trapping 119, 120
Apis cerana 33
ApiShield 116, 117
Apis mellifera 33, 129
Apis mellifera cypria 34
Arrival in Europe 47
Asian giant hornet 8
Asian Hornet Action Team - see AHATs
Asphyxiation 34
Association Action Anti Frelon Asiatique 109
Attaching a radio-tag 92
Attack 86
Attractant 118
Autumn trapping 120

B

Bait shadow 83
Bait station 70, 71, 72, 102
Bath 54
BBKA 54, 59, 60, 121
Bee ball 34
Bee carpet 33
Bee Tree 51
Belgium 48
Bendiocarb 96
Binoculars 71, 76, 86
Biocides 97
Biodiversity 39, 59, 101
Biological control 125, 128
Biology 11
Blot trap 104
Bordeaux 92, 106
Boston 55
Box trap 114, 122, 131
British Beekeepers' Association 54, 59, 60, 121
Brockenhurst 55
Brood 17, 21
Bumblebees 126
Bury 55
By-catch 104, 120

C

Camellias 11, 27, 52, 72, 95, 101
Camera trap 102
Canopy 82, 83
Cap 14, 21
Capture-mark-recapture 32
Carnivorous plants 127
Carrion 28, 118
Carrying capacity 22, 42
Castes 18
CEH 60, 65
Centre for Ecology and Hydrology 60
Channel Islands 48, 50
Cherry-picker 95, 100
Chickens 128, 131
Chimney 132
China 47
Climate change 49
Cocoon 14
Colony, Asian hornet
  activities 18
  destruction 95
Colony collapse (honey bee) 35
Colony losses (honey bee) 40, 120, 131
Colour coding 77
Comb 20, 21
Compass 71, 74, 76
Competition between Asian hornets 32, 42
Competition between hornet species 41

Confounding factors, flight times 79
Conops vesicularis 125
Cooling 17
Copulation 38
Cost of control 134
Cypermethrin 97

D

Dangerous nest locations 45
Dangers 16
Death, human 43, 44
Deformed wing virus 128
Defra 60, 65, 91
Denis le Jaffre 115
Destroying nests 62
Destroying primary nests 95
Destroying secondary nests 95
Detection rate for nests 23
Developmental stages 14, 21
  egg 14
  larva 14
  pupa 15
  timings 15
Devon Beekeepers' Association 66
Diatomaceous compounds 97
Diazinon 124
Diet, adult 27
Diet, larva 27, 28
Dispersal flights 39, 64
Disrupting hawking 131
Distance rule of thumb 76
Distribution, natural 47
Diversion traps 110
DNA technology 129
Dolichovespula saxonica 8
Dolichovespula media 8
Drone, machine 55, 56, 93, 98
Dungeness 56
Duster 96

E

Early males 16, 37
Ecologists 104, 120
Ecology 39
Ecosystems 62, 101
Effect on European hornets 41
Effects on humans 42

Egg 14
Electric harp 133, 134
Embryo nest 13
Entomopathogenic fungi 127
Entrance reduction 131
Eradication stage 61
Etiquette 70
European hornet 5, 16, 41, 114, 123
  nests 6
Europe, spread through 47, 48
Expansion of range 47
Exudate 50

F

FDGDON 106, 108
Feather, attaching 82, 84, 85
Feeding honey bees 134
Feeding station 72
Fertile males 36, 37
Ficam D 96
Finding Asian hornet nests 69
Fipronil 123, 124
First workers 14
Fish 30, 73, 101, 118
Fishmongers 28
Flight path 79, 82
Flight speed 26
Flight times 76, 79
  confounding factors 79
  haywire 83
  obstacles 79
  time at nest 79
Fluff, attaching 84
Food for adults 27
Food for larvae 27
Food traps 104
Foraging 18, 25, 26, 29
  range 25, 54, 70
Foraging paralysis 35, 134
Founding 12
Foundress 11, 104
Fowey 55
France 48
France Nature Environnement 108
FREDON 106, 108
French trapping data 104
Fungal spores 98

G

Genetic 'bottleneck' 129
Genetics 129
Geraniol 30
Germany 48
Giant wood wasp 6
Glossary 135
Glue trap 118
Google Maps 71, 76
Guernsey 50
Guildford 56
Gulff UV resin 84
Gynes 36, 37

H

Half-lives 124
Harberton 65
Harmonic radar 94
Harpe electrique 133
Hawking 31, 32, 35, 110, 119
Heat ball 34
Hibernation 11
Hibernation of the queens 39
Hissing 34
Hive defence 131
  chimney 132
  electric harp 133
  harpe electrique 133
  museliere 132
  muzzle 132, 134
  raquet 133
  vertical muzzle 132
Honey bee colony losses 40
Honey bee defence 33
  asphyxiation 34
  bee ball 34
  bee carpet 33
  heat ball 34
  hissing 34
  shimmering 33
  stinging 34
Honey bee feeding 36
Honey bee predation 30, 40, 111, 119
  apiary 31, 32
  colony collapse 35
  entering the hive 33
  foraging paralysis 35
  hawking 31, 32, 35, 110, 119, 131
  honey bee defence 33
  synergistic predation 41
  timings 31
  weather 31
Honey bee products 40
Honey buzzard 125
Hoovering 98
Hornet mimic hover-fly 7
Hornet suit 44
Horntail 6
Hotspot 24
Hover-fly 7
Hull 55

I

IAS 59
Identification 3, 121
  castes 9
  males 9, 37
  queens 10
  sexes 9
Identification posters 9
Infertility gene 130
Information sheet 73
Infra-red camera 55, 93
INRA 109
Insect decline 101
Insect energy harvesting 94
Insect growth regulators 124
Insecticide 95, 96, 97, 123
Institut technique et scientifique de
  l'abeille et de la pollinisation 107
Invasive alien species 39, 59
Israeli acute paralysis virus (IAPV) 40
Italy 40, 48
ITSAP 107, 109
IUCN 59

J

JABEPRODE 115
Jersey 50, 52, 53, 92
Jersey method 70

L

Lance 95
Landmarks 74

Landowner 70
Larva 14
Larval diet 27
Larval regurgitations 27
Late autumn/ winter trapping 103
Life cycle diagram 12
Life history 11
Life span 25
L'Institute national de la recherche
    agronomique 109
Liskeard 55
Location naming app
    What Three Words 87
Loire Valley 54
Longevity 25
Looking for the nest 82
Lorient 108
Loss of income 40

M

Males, diploid 54, 129
Males, early 16, 37
Males, fertile 36, 37
Mallorca 48, 98
Management scenario 64
Map of Europe 48
Mating 38
Meat ball 28
Meconium 14, 21
Mesoscutum 10
Migration 11, 39
Mites 128
MNHN 109
Monitoring traps 63, 101, 110, 113
Morbihan data 106
Mother queen 38
Moulting 14
Mouse guards 131
Museliere 132
Museum National d'Histoire Naturelle
    109
Muzzle 132, 134

N

Naïve native prey 40
Nantes 127
National Bee Unit 51, 59, 60, 65

National Contingency Plan 59
Native distribution of Asian hornet 47
NBU 60, 65, 69, 93, 97
NBU diagram 60
NBU monitoring trap 103, 114
NBU response to sighting 63
Nectar 27, 118
Nematodes 126
Neonicotinoids 124
Nest decline 39
Nests, Asian hornet 12
  cooling 17
  dangerous 45, 86
  destroying nests 62, 95
  destroying primary nests 95
  destroying secondary nests 95
  detection rate 23
  distribution 22
  entrance, embryo 13
  entrance, primary 18
  entrance, secondary 19
  finding 69, 93
  hotspots 24
  location 22, 23, 24, 45, 86
  looking for the nest 82
  nest decline 39
  nest density 22, 42, 45
  number of cells 25
  number of individuals 25
  papier-mâché 12
  primary nest 12
  pulp 12, 29
  relocation 18, 19
  secondary nest 19
  size 21
  thermoregulation 17
Netherlands, the 48
Neurotoxins 43
New Alresford 55
New Zealand 107, 123
NNSS 9, 60, 61
Non-native Species Secretariat 9, 60, 61
North American pitcher plant 127
Numbers of individuals 25

O

OPIE 108
Orientation flights 15

Orvis Egg Yarn 84
OS aerial map 71

P

Paintball gun 98
Papier-mâché 12
Parasites 125
Permethrin 97
Permission to go onto property 70
Permits 63, 69
Pernis apivorus 125
Petiole 13
Pheromermis vesparum 126
Pheromone trap 64, 104, 119, 122
Pie charts 29
Piezoelectric generation 94
Plastic bottle trap 112
Pnemolaelaps niutirani 128
Poisoning 64, 97, 123
Pollination 40, 42
Polyethism 18
Poole-Cherbourg ferry 55
Population dynamics 107
Portugal 48
POSCA (uni POSCA) marker pen 71, 75, 139
Post hibernation migration 11
Pottery 47
Predation 28, 29, 40
Prey items 29
 pie charts 29
Prey pellets 27, 29
Primary nest 12, 17, 70
Pupa 15
Pyrethrum 97

Q

Queen
 foundress 11
 gyne 36, 37, 101
 hibernation 39
 identification 10
 migration/dispersal 39
 mother queen 38
 queen colony phase 14
Queen catcher 71, 75

R

Radio-tags 90
Radio telemetry/tracking 55, 63, 90
Range, foraging 25
Raoul's cage 115
Raquet 133
Rate of spread 11
Receiver 92
Reduction of entrances 131
References 140
Relocation of nest 18, 19
Reporting a sighting of an Asian hornet 9
Resources 138
Rule of thumb, distance 76

S

Saliva 14
Sap 11, 27, 101, 118
Sarracenia 127
Satellite map 71
Scotland 55
Scouting 26
Secondary nest 19
 photos 87, 88, 89
Selection of honey bee strains 129
Sentinel hives 61, 110
Sex pheromones 38, 119
Sexual stages 36
 numbers of sexuals 37
Shimmering 33
Shrimps 73, 118
Sighting 9, 62, 69, 121
Size 3
Small hive beetle 61
Somerset, north 54
Spain 48, 123, 134
Speed of spread 47
Sphaerularia bombi 126
Sphaerularia vespae 126
Spread, Asian hornet 47
 climate 49
 Europe 48
 speed of spread 47
Spring trapping 102, 104, 105, 109, 119
Squirting venom 45
Steinernema feltiae 126

Sticky trap 117
Stings 42, 43
  alarm pheromone 44
  allergy 44
  attack 44
  in China 45
  size 44
  toxins 43
Stopwatch 71, 75
Stress for beekeepers 40
Stress for honey bees 35
Subspecies 3, 49
Sulphur dioxide 97
Summer/early autumn trapping 102
Summer trapping 102
Superorganism 40
Supplementary feeding 36, 134
Surveillance 63
Suterra wasp attractant 51, 72, 118
Switzerland 48
Synergistic predation 41

T

Temperature 11
Temporal polyethism 18
Tetbury 54, 57, 59
Thermoregulation 17
Tippex 75
Tours 16, 120
Tracking 72
  feather 82
  Jersey method 70
  tracking experience illustration 80
Traffic, Asian hornet 82, 86
Transhumance 134
Transmitters 90, 91
Trap design 110
Trapping 101, 131
  apiary 119
  Blot trap 104
  box trap 114
  by-catch 104, 119
  food traps 104
  glue trap 118
  killing trap 102
  late autumn/ winter trapping 103, 120
  monitoring traps 101, 110, 113
  monitoring without trapping 102
  Morbihan 106, 108
  spring trapping 102, 105, 119
  sticky trap 117
  summer/early autumn trapping 102, 119
  trapping at different times 119
  trap practicalities 110
  UNAF on trapping 108
Trapping bait graph 31
Trap practicalities 110
Trélissac 108
Triangulation 55, 78, 93
Trojan method 123, 134
Trophallaxis 27

U

UAV ('drone', machine) 55, 56, 93, 94, 98
UK incursions 54
UK, possible spread through 48, 54
UNAF on trapping 108
Uni Big Fly thread 84
Uni POSCA marker pen 71, 75, 139
University of Exeter 91
Urban environments 24, 28, 29, 42, 45, 64
Urocerus gigas 6
Usurpation 16, 107

V

Varroa destructor 40, 128
Venom 43, 44, 45, 118
Vespa affinis 16
Vespa analis 126
Vespa crabro 5, 19, 26, 41, 66, 128
  nests 6
Vespa mandarinia 8, 30
Vespa orientalis 34, 126
Vespa simillima 19, 126
Vespex® 123
Vespids 123
Vespula rufa 8
Vespula vulgaris 7
Véto-Pharma trap 102, 110
Volucella inanis 126
Volucella zonaria 7
Vertical muzzle 132

W

Walkie-talkies 71
Wasp attractant 30, 71
   Suterra wasp attractant 51
Wasps 7, 70, 72, 123
  Dolichovespula saxonica 8
  Dolichovespula media 8
  nests 7, 8
  Vespula rufa 8
  Vespula vulgaris 7
Water 13
What Three Words 87
Wildlife 95, 98
Wind 31, 79
Winter colony collapse 35
Winter colony losses 41
Winter trapping 103, 120
Woolacombe 55, 57, 65
Workers
  catching 74
  colour coding 77
  first workers 14
  marking 75
  numbers of workers 36

X

Xenos moutoni 126

The Bee's Knees!
Bee Suits, Hives and Supplies

Ultra Ventilated Bee Suit

Keep ultra cool with fully ventilated fabric
Stay ultra protected with 5mm thick sting protection
Superior comfort which will keep you cool all day

Quality assured

Made to very high standards in the UK
Developed and extensively tested by professionals
Field tested with bees, wasps, Asian & EU hornets
Particularly robust veiling made to our specification
No reported stings since manufacture in early 2018
Made with highest quality materials and components

www.bbwear.co.uk

t. +44 (0)1872 562731   e. shop@bbwear.co.uk

Lightning Source UK Ltd.
Milton Keynes UK
UKHW051010090519
342373UK00003B/22/P